T0258010

Psychedelic Chemistry

by Michael Valentine Smith

Psychedelic Chemistry

by Michael Valentine Smith

RONIN
Berkeley, CA
www.roninpub.com

Psychedelic Chemistry

Copyright 1981 by Michael Valentine Smith

ISBN: 978-1-57951-190-6

Published by

RONIN Publishing, Inc.

PO Box 3436

Oakland, CA 94609

www.roninpub.com

Production:

> Cover Design: Beverly A. Potter.
> Drawings: Laurel D. Maroita

Library of Congress Card Number: 73-79212

Distributed to the book trade by PGW/Perseus

To Albert Hoffman and Ludwig Wittgenstein,
who opened the doors of perception.

I shuttered as I took note of the strange things that were going on inside me. An exquisite pleasure had invaded me.... Immediately it had made the vicissitudes of life indifferent, its disasters inoffensive, its brevity illusory—in much the same way as love operates, filling me with a precious essence: or rather this essence was not in me, it was me. I had ceased to feel mediocre, contingent, or mortal.

—Marcel Proust

What's made Milwaukee famous has made a loser out of me.

—Jerry Lee Lewis

Them that dies is the lucky ones.

—Long John Silver

A little poison now and then: That makes for agreeable dreams. And much poison in the end, for an agreeable death.

—Friedrich Nietzsche

They are made for life, not for thought. Yes, and he who thinks, what's more, he who makes thought his business, he may go far in it, but he has bartered the solid earth for the water all the same, and one day he will drown.

—Herman Hesse

No one's mouth is big enough to utter the whole thing.

—Alan Watts

Reason craps out in an instant when it is out of its safe, narrow bounds.

—Don Juan

FREE PRESS

This book is being reissued, without change, as previously published by Loompanics Unlimited under the freedom of the press guaranteed by the First Amendment of U.S. Constitution. The information in this book is presented for lawful purposes only, such as for researchers, scientists, novelists, and others needing it. Possession of chemicals and carrying out procedures described herein may be illegal. Readers should consult an attorney and check state and federal laws before acquiring any chemicals or carrying out any procedures described in this book. The publisher and author do not advocate breaking any laws or engaging in any risky experiments. Consult with a licensed chemist before experimenting with any procedure decribed.

WARNING TO READER:

This book is not for children or people who are not trained chemists. The procedures described in this book require a thorough knowledge of advanced lab techniques and a safe, clean laboratory and should not be undertaken by amateurs. Inexpert procedures can asphyxiate you, blow you up, set you or your house on fire. If the end product is imperfectly prepared, you and your friends can be poisoned. The publisher does not warrant the efficacy of any procedures described in this book.

NOTE:

Orders for certain chemicals used to make psychedelics—especially large orders in suspect areas—are sometimes checked by narcs. Indole, lithium aluminum hydride, trimethoxybenzaldehyde, phenyl-2-propanol, diethylamine, olivetol and ergotamine are among those watched. The vast majority of the homologs and analogs described are, however, legal to manufacture and use. A list of DEA watched chemicals is in the back of this book.

For lists of federally proscribed drugs, see the *Code of Federal Regulations 21:* part 300 to end, the *U.S. Pharmacopoeia 19:690* or write to the Drug Enforcement Administraiton, Dept of Justice, 1405 1st St. N.W., Washington, D.C. 20537. There are also state laws to contend with.

PREFACE TO THE FIRST EDITION

The whole field of psychedelics, including areas of botany, chemistry, and pharmacology, is still in a primitive state. Thousands of potential psychedelics have been synthesized which have not been tested on man, some of the more promising of which are indicated in these pages. Also, anyone conversant with contemporary advances in synthetic methods could devise better ways to synthesize most psychedelics. I have endeavored to gather here all the more useful information on the synthesis and structure activity relationships of the compounds loosely referred to as psychedelics, of which LSD, mescaline, and the active constituents of *Cannabis* are the most notorious prototypes. In each section, the simplest methods, or those giving the highest yields, are given first. The many synthetic routes contained in the literature, but omitted here, will usually be found to involve greater difficulty, or lower yield. Each synthesis is an abbreviated, reworded, and often translated extract from a longer paper. While I have tried to make them accurate and coherent, a fuller understanding as well as a correction of the inevitable blunders may be achieved by consulting the original papers cited. Occasionally, material will be found which is not contained in the cited work, and which is my attempt to supplement the description.

Though possessing little knowledge of chemistry, I have undertaken this task because the only previous efforts of which I am aware are so dismally inadequate. It is my fondest hope that some highly skilled chemist will use the present effort as an outline in producing a thoroughly competent work on psychedelic chemistry. Such a document should significantly accelerate the psychedelic revolution.

PREFACE TO THE SECOND EDITION

The scenario for the psychedelic revolution was completed in the late sixties and nothing surprising has emerged since. Though the vast majority of substances discussed here are clearly relatively harmless and have tremendous potential for medical and psychological therapies as well as for facilitating personal growth, there has been only a feeble trickle of official research on them. This is due to the anti-drug laws, coupled with the cowardice and stupidity of officialdom that is almost invariable. One exception is Grof's superb book delimiting the parameters of psychedelic psychotherapy (REALMS OF THE HUMAN UNCONSCIOUS), but for the most part the field has been abandoned to unofficial research.

I have included much new material that has come to my attention since the first edition of this book, but with the field of organic chemistry growing as it is, there are undoubtedly syntheses that have eluded me. I would appreciate readers calling attention to these.

I am indebted to Gyna Parthenos and her manuscript A NEW METHOD OF LSD PRODUCTION (unpublished) for some of the intriguing new material in the LSD chapter.

TABLE OF CONTENTS

GLOSSARY

cyclize - to form into a circle. Specifically, to link the OH group of CBD to the carbon atom to form the 3 ring compound THC, from the 2 ring compound CBD.

decarboxylate - to remove a carboxyl (-COOH) group from a molecule. Specifically, to remove the carboxyl group from THC acid or CBD acid, by heating, to give THC or CBD.

isomerize - to rearrange the atoms of a molecule such that the molecular weight remains the same, but the chemical or physical properties change. Specifically, when CBD is cyclized to THC, it has undergone isomerization.

polymers - a complex chemical compound formed of many simpler units. Cellulose is a sugar polymer and gelatin is an amino acid polymer.

synergize - the working together of two forces or chemicals to produce an effect greater than the sum of their individual effects. Specifically, THC and CBD together may synergize to produce redder eyes than the sum of the redness of the same amounts of the two, taken alone.

PSYCHEDELICS AND SOCIETY

Primitive man likely began altering his consciousness with psychedelics very soon after he first evolved into a creature that we could term human. Virtually every area of the world possesses some plants with psychedelic properties, and man seems to have a remarkable ability to find just which roots and leaves are most effective in blowing his mind. Unfortunately, Western Europe and North America contain few psychedelic plants, and even though primitive man may have brought marijuana seeds and the opium poppy into these regions, he would have found them to be poor psychedelics when grown in a cold, wet climate. In any case, psychedelic use largely disappeared in Western societies and did not reappear for several thousand years.

Man has undoubtedly suffered considerably from the abuse of various consciousness altering drugs, in particular alcohol and the opiates; but less well known, at least in the West, are the benefits which psychedelics have produced. It is gradually becoming evident that the world owes much of its early art, music, literature, indeed perhaps the very fabric of many ancient cultures directly to the effects of the psychedelic state. Many millions of Westerners have recently gained personal experience with the euphoria, aphrodisia, relaxation, and stimulation of creative imagination which can result from psychedelic use. The failure of many otherwise enlightened people to accept psychedelics has many causes, one of the more important of which is probably that they are often among the fortunate few to escape the neuroses, psychoses, abuse of amphetamines, tranquilizers, barbituates, and alcohol, which is the fate of the majority of the straight world.

While there are good reasons for supposing that psychedelics have much to offer society and at worst have considerably less potential for doing damage than many legal drugs such as tobacco and alcohol, an adequate analysis of the total impact of psychedelics will not be possible for at least several decades and perhaps not until they are legal. Legality could come about very soon in some areas, since a town can pass laws against manufacture, possession and sale which provide very light penalties (e.g., a $1 fine) for conviction.

Presumably, anyone busted could be prosecuted under city laws and thus double jeopardy might prevent state or federal laws from being effective.

Even though such laws will eventually be declared unconstitutional because of conflict with state or federal laws, others could be passed, and there are other legal grounds for challenging drug laws.

For a useful article on ways of cross-examining the official chemist in drug cases, see Contemporary Drug Problems 2,225 (1973), and Microchemistry 4:555-67 (1960).

Drug Damage

For virtually every drug there is a portion of the population which will have an undesirable response. Medical journals are replete with descriptions of drug damage and fatalities. In the US, penicillin and aspirin each account for some 200 deaths yearly, and it was recently estimated that prescription drugs are implicated in the deaths of some 29,000 Americans a year. The most toxic drug commonly ingested in large amounts is probably nicotine; about three cigarettes contain a lethal dose (although some brands now require two packs or more for a lethal dose). In contrast, the psychedelics seem to be relatively safe and non-toxic. Some people, perhaps one in a hundred, can develop a temporary psychosis which rarely lasts more than a few hours or days, and which generally seems to indicate pre-existing psychopathology. In a more enlightened future, psychedelics may be universally applied to expose latent mental illness. There is no sound evidence that any psychedelic damages human brains, genes, embryos, etc. *Of course, all drugs are best avoided in the early months of pregnancy, and excessive or prolonged use may have undesirable effects.*

Psychedelics may cause subtle psychological changes leading to apathy, fuzzy thinking, paranoia, etc., but the many studies noting such effects in users are almost all worthless, since they lack adequate controls. Psychiatric complications are common side effects of many prescription drugs (e.g., see "Psychiatric Complications of Medical Drugs," R. Shaden (Ed.), 1972); and Heusghem and Lechat "Les Effets Indesireables des Medicaments" (1973).

Opiates (Heroin, Demerol, morphine, opium, etc.), non-hallucinogenic amphetamines (Methedrine, Preludin, "bennies," etc.) and barbituates ("reds," "yellows," etc.) are all addictive, lend themselves to intravenous injection, and when used heavily usually lead to bodily damage and frequently to death. Intravenous injection of anything under unsterile conditions is a bad idea, since it will almost invariably lead eventually to tetanus, hepatitis, bacterial or fungal infection of the heart and arteries, partial paralysis, etc. The death rate among intravenous addicts is extremely high (a recent English study found the rate for smack heads thirty times higher than for a control group) and the best that can be said for these drugs is that they are a rather expensive and unreliable way of committing suicide.

Medical Use of Psychedelics

It is often said that psychedelics have no recognized medical use. Anyone who examines the technical literature with even a modicum of critical competence realizes that this is true simply because there has been virtually no adequate research. Psychedelics clearly have tremendous potential in medicine (e.g., psychotherapy, antidepressants, appetite stimulators, analgesics, aphrodisiacs, etc.) as well as in biology and psychology. Psychotherapy is the most obvious area of application, and though many studies have been done, very few deserved publication. Careful selection of subjects, adequate controls, and careful followups are uncommon, and the techniques used usually border on the idiotic. For example, the use of LSD in the treatment of alcoholism: Four different studies reported in 1969 found, in contrast to other work, that LSD was of no use in the treatment of alcoholism. These four studies shared the following characteristics: 1) there was little or no preparation for the drug experience, and a large dose was given the first time; 2) the drug was given in a hospital setting; 3) the patient had to trip alone, and had no one present whom he loved or trusted; 4) there was little or no effort to use psychotherapy before, during, or after the drug experience; 5) perhaps the most important, the LSD was given only once. Since all five of these conditions are contrary to what experience has shown to be the most effective ways of using psychedelics, the negative results of the studies are hardly surprising. To varying degrees, such inadequacies are present in most medical research with psychedelics, and progress in this area can be expected to be very slow, especially in view of the legal hindrances due to neanderthaloid legislators.

For a recent discussion of the potential value of LSD psychotherapy and the relative lack of adverse side effects, see Psych. Bulletin 79,341(1973). Above all, see Stanislav Grof's definitive study REALMS OF THE HUMAN UNCONSCIOUS (1976).

Synthetic vs. Organic

Many people believe that organic or natural psychedelics such as peyote, magic mushrooms and marijuana are safer or produce better trips than synthetic compounds. This is almost certainly false, since any plant material contains hundreds of compounds, many of which have a definite toxicity, but few of which have psychedelic properties (they tend to make you sick, not stoned). The various impurities or the additives (e.g., amphetamine, belladona, strychnine) sometimes found in synthetic preparations are probably no more toxic than many of the compounds found in the psychedelic plants, and like these compounds, such additives or impurities probably have relatively little effect on the trip.

There is a great deal of superstition regarding purification of psychedelics. Actually, any impurities which may be present as a result of synthetic procedures will almost certainly be without any effect on the trip. If there are 200 micrograms of LSD in a tablet, there could only be 200 micrograms of impurities present even if the LSD was originally only 50% pure (assuming nothing else has been added), and few compounds will produce a significant effect until a hundred to a thousand times this amount has been ingested. Even mescaline, which has a rather specific psychedelic effect, requires about a thousand times this amount.

It is possible that iso-LSD may block LSD effects somewhat and inhibit the cosmic trips that can result from high doses; this is however unproven. Nevertheless, the prime reasons for a lack of cosmicity are undoubtedly low doses and the development of tolerance. A single exposure to LSD or other psychedelics may produce an adaptation or tolerance that lasts the rest of your life (seeing the ocean for the first time is not a repeatable experience). Furthermore, as seems to be the case with the active chemical (THC) and its inhibitor (CBD) in marijuana, the presence of the inhibitor may sometimes result in a more pleasant experience. Only careful studies in which varying amounts of iso-LSD are added to LSD will decide the issue.

Trip Differences

If a psychedelic is taken several days in succession, some degree of tolerance (failure to produce a trip) develops. If a different psychedelic is then taken and this also fails to produce a trip, the two compounds are said to produce cross tolerance, which strongly indicates that they act in the same way and create roughly the same kind of trip. LSD, mescaline, and psilocybin (and probably the hallucinogenic amphetamines) all produce cross tolerance, and there are some studies which indicate that people are unable to tell them apart. In comparing trips, it should be kept in mind that mescaline has seldom and psilocybin very rarely been available on the black market. Virtually all "psilocybin," as well as most of the "mescaline" has been LSD or one of the hallucinogenic amphetamines (when they haven't been atropine, PCP, speed, etc.)

PharmChem Dept. AA, 3925 Bohannon Drive, Menlo Park, CA 94025, will qualitatively analyze any sample mailed in for $10 and will give results by phone (415-328-6200) five days after the sample is received. Anonymity can be achieved by assigning an arbitrary 6 digit number to your sample and giving only this number when you call. They would like to know what you think your sample is, its street price and general area of origin. Envelopes with significant size specimens should be marked "Hand Cancel". The feds won't let anyone give out quantitative results anymore. PharmChem desires that letters be marked "Hand Cancel" and that you allow 5 days after receipt for results. Their rates are now $10 for a qualitative and $15 for a quantitative analysis (for the latter, you must first get from them a controlled drug transfer form to send with the sample and the $15).

Perhaps the only reliable way to identify a psilocybin trip is by its short duration; most trips are completely over in six hours or less. THC, DMT, glycolate esters and very likely muscimole probably do not produce cross tolerance with each other or with the LSD-mescaline-psilocybin group, as would be expected from the distinct kinds of trips produced by each of the former compounds. Other than the synthesis of new compounds, the most fertile source of new trips lies in the combination of varying amounts of known psychedelics.

Although tranquilizers tend to inhibit the effects of psychedelics if taken shortly before or during a trip, pretreatment (several hours to several days) with a tranquilizer will often enhance the effects. This enhancement is highly variable depending on the type and amount of tranquilizer and psychedelic, time between ingestion, etc. Prior administration of some tranquilizers is also useful in combatting the nausea which, though transient, is a common unpleasant side effect of most psychedelics. Certain phenothiazines (Stelazine, Compazine, Prolixin, Vesprin, Trilafon) are most effective as antiemetics. Immediate relief of nausea may be obtained from various nonprescription products, of which Emertrol is perhaps the best.

6

With most psychedelics, their activity can probably be considerably enhanced by prior (or possibly concomitant) use of a monoamine oxidase inhibitor (e.g., isocarboxazid (Marplan), nialamide (Niamid), phenelzine (Nardil), and tranylcypromine (Parnate)). Some compounds (e.g., DMT) which have no oral activity, can probably become orally active. These compounds are often prescribed as antidepressants, but it is not a good idea to use them frequently or in large doses. For antidotes to the hallucinogens see Amer. J. Hosp. Pharm. 30,80(1973).

Useful Books

THE ALKALOIDS, R. Manske and S. Holmes (eds). 16 volumes, 1951-76.

CHEMISTRY OF THE ALKALOIDS, S. Pelletier (ed), 1970.

ETHNOPHARMACOLOGIC SEARCH FOR PSYCHOACTIVE DRUGS, D. Efron (ed.), 1967.

THE HALLUCINOGENS, A. Hoffer and H. Osmond, 1967.

MARIJUANA - THE NEW PROHIBITION, J. Kaplan (1970). (The only competent study of legalization)

PSYCHOPHARMACOLOGICAL AGENTS, Vol. 1, M. Gordon (ed.), 1964; Vol. 4, 1976.

PSYCHOTROPIC DRUGS AND RELATED COMPOUNDS, E. Usdin and D. Efron, 1967, and supplement.

SOMA: DIVINE MUSHROOM OF IMMORTALITY, R. Wasson, 1968.

THE NATURAL MIND, A. Weil, 1972.

FLESH OF THE GODS, P. Furst (ed.), 1972.

THE BOTANY AND CHEMISTRY OF HALLUCINOGENS, R. Schultes and A. Hofmann, 1973.

NARCOTIC PLANTS, W. Emboden, 1972.

MARIJUANA:
THC and ANALOGS

History

Cannabis sativa has been a cherished friend of an ever-increasing proportion of mankind since prehistoric times. The genus Cannabis contains only this one species, but its appearance and psychedelic properties vary tremendously depending on growth conditions. It seems to have originated in Asia, but its "medicinal" properties and the long fibers in its stem used to make rope (hemp) have led to its being carried all over the world. The early American colonists brought seeds with them and George Washington, "the father of our country," was among its cultivators.

There is evidence for Cannabis use as early as 5,000 B.C. in Central Europe and there are probable references to it in the earliest writings of China and Egypt. The frozen tombs of the inhabitants of Siberia in about 500 B.C. have yielded quantities of seeds along with a variety of devices for burning them. A very hot, dry climate appears necessary for producing grass with high psychedelic activity, which may account for the failure to use Cannabis as an intoxicant in Northwestern Europe.

Cultivation

Grass seems to grow best in loose, well fertilized soil (manure, superphosphates). The soil should be well tilled, and the seeds (actually drupes or achenes, a type of nut) sown about one inch apart in rows about one foot apart and covered with about ¾ inch of soil. Some say it is best to use dark (not green) seeds, and to soak them overnight in water before sowing. They can be sown in flatboxes and transplanted when about two to six inches high (about two weeks). After the plants have their first two leaves they can be thinned by pulling up those which look the weakest.

Cannabis is usually dioecious (male and female flowers on separate plants). The female plants are widely believed to contain more THC than the males, but this varies from strain to strain. In India, the males are apparently weeded out before they have a chance to pollinate the females and initiate seed development in the belief that the unfertilized females produce stronger grass. This is unproven, and even if true, the total yield of THC from a field of males plus fertilized females may well be greater. The males are generally taller and thinner until almost mature (3 - 4 months), when the females may become taller. Males tend to yellow and die some weeks before the females. Harvest takes place around September in Europe and Japan and around July in North Africa. The sex of the plants can be influenced by soil conditions, some experiments giving more males in moist, heavily manured soils. Long light periods tend to masculinize, whereas plants grown under short day conditions (for example, those seeded in northern latitudes in winter) tend to produce intersexual flowers, males changing to females; a condition also resulting from debudding males. It has recently been found that male flowers tend to change to females after early spraying with Ethrel (2-choloro-ethane-phosphonic acid). Under some conditions, flowering may occur in as little as two weeks. A sizable portion of the leaves can be harvested after about two months and the plants will continue to grow. The effects of these various manipulations on the THC yield of a single plant, or the crop as a whole, have yet to be determined. "The Induction of Flowering," L. Evans (ed.), 205(1969) is a good reference on *Cannabis* flowering.

Pinching off the shoot tip when the plants are just beyond the seedling stage, and pinching off subsequently developing side shoots at about weekly intervals thereafter in order to keep the plant only a few feet tall, can produce plants so altered in shape, color and leaf morphology as to be unrecognizable as cannabis. The resins are said to accumulate on the surface of such distorted plants to such an extent that it looks crystalline, and the resulting grass is supposed to be as strong as hashish.

For definitive references on the cultivation and chemistry of marijuana, see MARIJUANA GROWERS GUIDE, And/Or Press (1978) and MARIJUANA POTENCY, And/Or Press (1977).

THC Extraction

Some samples of grass have considerable THC acid. In order to extract this it is first necessary to decarboxylate it by heating it to 110° for fifteen minutes. Grass grown commercially for its fiber content, as well as that grown in northern Europe or much of the U.S.A., contains mostly the inactive cannabidiol and cannabidiolic acid. These compounds can be extracted and used to synthesize the

9

active THC and THC acid (by smoking, not active orally) in one step. See TET 21,1228(1965) or the following. To extract, add 50g grass/liter petroleum ether or benzene; soak twelve hours with occasional shaking; filter and extract petroleum ether three times with a solution containing 5% NaOH and 5% Na_2SO_3. Acidify the aqueous extracts with cold dilute sulfuric acid and extract with ether or chloroform which is dried, evaporated in a vacuum to yield the cannabidiolic acid. THC and cannabidiol remain in the petroleum ether which can be dried and evaporated in a vacuum and the residue added to grass. THC acid is converted to THC by boiling in benzene for seven hours.

THC Extraction for Smuggling or Converting Poor Grass to Good

This is recommended to anyone who wants to smuggle or otherwise conceal grass or hash. One hundred kilos of grass will convert to about two or three kilos of resin, which can later be redissolved and absorbed on alfalfa, etc. See *Lloydia* 33,453(1970) for a method of large scale extraction.

Cover grass or hash with methanol, benzene, petroleum ether, or isopropyl (rubbing) alcohol. Allow to soak for about twelve hours, filter and repeat soaking with fresh solvent. This process can be speeded up by gently heating the solvent plus grass (no flames) for about three hours, adding more solvent if necessary. Evaporate the combined solvent extracts until a resin is obtained or until syrupy and absorb the syrup on grass, etc. Repeat the process until no more resin is obtained, or until smoking some of the residual grass indicates that all the THC has been extracted. Methylene chloride works very well. Unleaded gas, preferably without additives (white gas), paint thinner or remover, or turpentine should be satisfactory. For a detailed discussion, see MARIJUANA POTENCY, And/Or Press (1977).

Dosage

Smoking seems to destroy most of the THC, but even so, this is several times more efficient than eating it, since the smoking dose of Δ^1 THC is about 2mg and the eating dose about 10mg. Based on a 5% THC content, 1g of hash efficiently used should (by smoking) stone about 25 people.

Partial tolerance to THC develops rapidly and most users observe that they get much higher the first time a given batch of grass is smoked than on subsequent occasions. However, for unknown reasons, tolerance to grass of different origin seems less, leading some people to desire a different batch for each day of the week.

Official THC Tests

One-tenth gram powdered grass, 5ml petroleum ether; grind in mortar and let soak fifteen minutes. Filter and add 1ml of the petroleum ether carefully to 2 ml 15% HCl in absolute ethanol. Red color at boundary of two layers indicates THC. After shaking, the upper layer is colorless and the lower orange-pink which turns colorless after addition of 1 ml of water. Alternatively, evaporate the petroleum ether, add 2 ml Duquenois reagent (12 drops acetaldehyde, 1 g vanillin, 50 ml ethanol), 2 ml HCl and stir. Let stand 10 minutes and add 2 ml chloroform; shake and let separate. A purple chloroform layer indicates THC.

Also useful but less specific is the fact that THC gives a purple color with 5% KOH in ethanol. A few grains of sucrose will often intensify color development in these tests. (See *Bulletin on Narcotics* 22,33(1970)).

THC Chemistry

Δ^3-THC
($\Delta^{6a,10a}$-THC)

1",2"-Dimethylheptyl
homolog of Δ^3-THC(DMHP)

Hexyl homolog of Δ^3-THC
(synhexyl, pyrahexyl, parahexyl)

1"-Methyloctyl homolog
of Δ^3-THC (MOP)

1",2"-Dimethylheptyl
homolog of Δ^1-THC

Some widely tested synthetic cannabinoids

11

Is It Legal?

The Δ^1 and $\Delta^{1(6)}$ THC's with the n-pentyl in the 5' position (obtained by using olivetol in the syntheses) are naturally occurring and hence illegal, but the Δ^3 THC's and the numerous isomers, homologs and analogs of the Δ^1 and $\Delta^{1(6)}$ compounds are probably legal.

Apparently, recent federal legislation outlaws delta-1, delta-1(6), delta-3,4-THC's, both cis and trans and D and L and compounds. This still leaves hundreds of legal cannabinoids.

Structure-Activity Relationships

THC refers to tetrahydrocannabinol, and Δ refers to the position of the double bond. Various numbering systems are used, so the following equivalences should be noted: Δ^1THC = $\Delta^1$3,4-trans-THC = Δ^9THC and Δ^6THC = $\Delta^{1(6)}$THC = Δ^8THC = Δ^6-3,4-trans-THC.

Little careful human testing has been done, so data given here and elsewhere on the relative psychedelic activity of various cannabinoids is often only a rough guess. Δ^1THC and Δ^6THC have about the same activity which is about five times that of Δ^3THC. Cannabidiol, cannabidiolic acid, cannabinol, cannabigerol and cannabichromene all have very little or no activity. Only the ℓ (-) isomer of THC seems to be active. When the n-pentyl at the 5' position is replaced by 1,2-dimethylheptyl, potency and duration of action increases about five times, giving the most active THC analog yet tested.

It should be noted that recent testing has indicated that a 1,1-dimethylheptyl or 1-methyloctyl and probably similar side chains give THC's of equal or greater activity than the 1,2-dimethylheptyl cpd. However, the difficulty of synthesizing these compounds plus their very long action (up to several days or more) makes it doubtful whether they deserve all the interest they have generated among psychedelic enthusiasts. More concern should be devoted to the shorter side chains, since they would presumably allow one to get very stoned but to be straight again within a few hours, thus allowing the drug to be more easily manipulated.

Substituting N, O, or S atoms at various places or saturating the double bond to produce hexahydrocannabinol probably retains activity. (See CA 74,125667(1971) for S analogs.) Alkoxy side chains at 5' retain activity. Unsaturated side chains are as active as saturated ones. Ether moieties at the 5' position, but not at the 3', retain activity. Activity is retained if an additional alkyl is placed at 4′ but lost if placed at 6′. Activity is greatly decreased or lost if the H at the 4′ or 6′ positions is replaced by carboxyl, carbomethoxyl, acetyl or acetoxyl; if the hydroxyl is replaced by H; if the OH is at 5′ and the side chain at 4′. Methyl and/or ethyl at 1 and 5 retains activity, as does removal of the methyl at 1. An hydroxyl in the side chain is active, but not on the first carbon of the side chain. Esterifying the OH retains activity, but etherifying eliminates activity.

THC can be synthesized via cannabigerol and cannabichromene in low yield (TET 24,4830(1968), TL 5349,5353(1969), Proc. Chem. Soc. 82,(1964)). For several moderately difficult routes leading to $\Delta^1(^6)$ THC via cannabinol in about 10% yield, see LAC 685,126(1965). For a synthesis of $\Delta^1(^6)$ THC from cinnamyl derivatives and isoprene see JACS 89,4551(1967). A rather difficult synthesis of Δ^1 and $\Delta^1(^6)$ THC is given in JACS 89,5934(1967). For a variety of THC analogs of unknown activity see BSC 1374, 1384(1968); JCS 952(1949); JACS 63,1971, 1977,2766(1941), 64,694,2031,2653(1942), 67,1534(1945), 70,662(1948), 71,1624(1949), 82,5198(1960); CA 75,48910 (1971); TL 3405(1967); JMC 11,377(1968); CT 2,167(1967); CA 76,126783(1972).

Since 0 or 1 and perhaps 2 double bonds anywhere in the lefthand ring below, as well as changes in the size and position of the alkyl groups will probably all produce compounds with THC activity, many compounds similar to menthadieneol, menthatriene, verbenol, epoxycarene, pulegone and 4-carbethoxy-1-Me-3-cyclohexanone can be used in the methods below to get active THC analogs (e.g., isopipertinol will work (TL 945 (1972))). Also, 5-chlororesorcinol and 5-methylresorcinol (orcinol) have been shown to give weakly active THC's (see CA 76,33946(1972), US Patent 3,028,410 (1962), and TET 23,3435(1967) for syntheses of orcinol and related compounds). Unfortunately, recent data indicate that orcinol gives a THC with very low activity. It appears that delta-5 and delta-7 THC have very little activity. If the methyl groups at carbon 8 in THC are changed to longer alkyl groups, the activity decreases, but the replacement of the alkyl groups by hydrogen or other groups has not been carried out. Open chain analogs also have activity (see CT 2,167(1967)).

For new information on the structure-activity relationships of cannabinoids see JMC 16,1200(1973), Arzneim, Forsch 22,1995

(1972), and Chem. Revs. 76,75(1976).

For THC analogs see JMC 19,445-71,549-53(1976); Eur. JMC 10,79(1975); Phytochem. 14,213(1975); CA 82,57564,170672-3(1975); Diss. Abst. 34B,1442(1973); J. Labelled Cpds. 11,551 (1975); Compt. Rend. Acad. Sci. 281C,197(1975).

For THC in one step from chrysanthenol see Experientia 31,16(1975).

Δ^8-THC Δ^1-THC

Two commonly used numbering systems for the same molecule

The following gives the synthesis of a water soluble THC derivative which is equipotent with THC and perhaps more rapidly acting (see Science 177,442(1972)). Stir equimolar amounts of THC, dicyclohexylcarbodiimide and gamma-morpholinobutyric acid hydrochloride (or gamma-piperidinobutyric acid hydrochloride) (JACS 83,2891(1961)) in methylene chloride at room temperature for 16 hours and filter, evaporate in vacuum (can triturate with ether and filter). The cost of synthetic THC will vary greatly depending on many factors, but high quality grass can probably be produced for under $20 a kilo.

For good reviews of marijuana chemistry see Prog. Chem. Natural Prod. 25,175(1967), Science 168,1159(1970), C. Joyce and J. Curry (Eds.), *Botany and Chemistry of Cannabis* (1970), JACS 93,217(1971), JPS 60,1433(1971), Ann. N.Y. Acad. Sci. 191(1971), Prog. Org. Chem. 8,78(1973), Marijuana-R. Mechoulam (Ed.) (1973), and Chemical Reviews 76,75(1976).

Only the first three methods below give the natural ℓ *(-) isomer of THC. The other methods give the racemic produce and consequently their yields of active THC are actually one-half that indicated.*

Syntheses of THC and Analogs

ℓ(-)- $\Delta^{1(6)}$ THC HCA 52,1123(1969), cf. JACS 96,5860

(1974).

Method 1

This method gives about 50% yield for THC and about 90% for the 1′,1′-dimethylpentyl analog.

Olivetol 4.74 g (or equimolar amount of analog), 4.03 g (+) cis or trans p-methadien (2,8)-ol-1 (the racemic compound can be used but yield will be one-half), 0.8 g p-toluenesulfonic acid in 250 ml benzene; reflux two hours (or use 0.004 Moles trifluoracetic acid and reflux five hours). Cool, add ether, wash with $NaHCO_3$ and dry, evaporate in vacuum to get about 9 g of mixture (can chromatograph on 350 g silica gel-benzene elutes the THC; benzene: ether 98:2 elutes an inactive product; then benzene: ether 1:1 elutes unreacted olivetol; evaporate in vacuum to recover olivetol).

Method 2

Dissolve the olivetol or analog and p-menthadienol or p-methatriene (1,5,8) in 8 ml liquid SO_2 in a bomb and fuse 70 hours at room temperature. Proceed as above to get about 20% yield.

$\mathcal{l}$(-)- ▲^{1}(6) THC JACS 89,4552(1967), JCS (C) 579(1971), cf. Diss. Abst. 35B,3843(1974), and Phytochemistry 14,213 (1975).

Convert (-) alpha-pinene to (-) verbenol (see precursors section). Add 1M(-) verbenol (racemic verbenol will give one-half yield), 1M olivetol or analog with methylene chloride as solvent. Add BF_3 etherate and let stand at room temperature one-half hour to get approximately 35% yield after evaporating in vacuum or purifying as above to recover unreacted olivetol. Solvent and catalyst used in Method 1 above will probably also work. Either cis or trans verbenol can be used. The JCS paper adds 1 g BF_3-etherate to a solution of 1 g olivetol and 1.1 g verbenol in 200 ml methylene chloride and let stand two hours at room temperature. JACS 94,6164(1972) recommends two hours at -10° C, then one-half hour at room temperature and the use of cis rather than trans verbenol (the latter gradually decomposes at room temperature). The reaction is also carried out under nitrogen, using twice as much verbenol as olivetol, 0.85 ml BF_3 etherate and 85 ml methylene chloride/g verbenol (both freshly distilled over calcium hydride) to give ca. 50% yield. See also JACS 94,6159(1972) for the use of citral and Arzneim. Forsch. 22,1995(1972) for use of p-TSA.

In the synthesis of THC with verbenol, the cis isomer is preferable to the trans since the latter decomposes at room temperature. Pinene or carvone give active THC's (JMC 17,287(74)).

Method 3

ℓ (-)-$\mathbf{\Delta}^1$ and $\mathbf{\Delta}^{1(6)}$ THC JACS 92,6061(1970), U.S. Patent 3,734,930.

1M (+)-trans-2-carene oxide (2-epoxycarene), 1M olivetol or analog, 0.05 M p-toluenesulfonic acid in 10L benzene; reflux two hours and evaporate in vacuum (or can separate the unreacted olivetol as above) to get about 30% yield THC. Olivetol can also be separated as described below. For synthesis of 2-epoxycarene ($\mathbf{\Delta}^4$ carene oxide) from $\mathbf{\Delta}^4$ carene (preparation given later) see p-methadieneol preparation (Method 2). 3-carene oxide gives 20% yield of $\mathbf{\Delta}^{1(6)}$ THC.

Methods for Racemic THC

Δ^3 THC JACS 63,2211(1941)

1M pulegone, 1M olivetol or analog, 0.3 M POCl$_3$, reflux four hours in 1 L benzene and evaporate in vacuum or pour into excess saturated NaHCO$_3$ and extract with dilute NaOH to recover unreacted olivetol. Dry, and evaporate in vacuum the benzene layer to get the THC.

$\mathbf{\Delta}^{1(6)}$ THC from Cannabidiol HCA 52,1123(1969)

Reflux 1g cannabidiol, 60 mg p-toluenesulfonic acid (or 0.003 M trifluoroacetic acid) in 50 ml benzene for 1½ hours. Evaporate in vacuum to get about 0.7 g THC. Alternatively, add 1.8g cannabidiol to 100 ml 0.005N HCl and reflux four hours. Proceed as above to get about 0.5 g THC (cf. JACS 94,6159 (1972)). *Nitrogen Analogs of* $\mathbf{\Delta}^3$ THC CA 72,66922(1970); JACS 88,3664(1966), TL 545(1972).

5.4 g olivetol or 0.03M analog, 5.8 g 4-carbethoxy-N-benzyl-3-piperidone hydrochloride or 0.03M analog (JACS 71,896(1949) and 55,1239(1933) give an old and clumsy synthesis, and Heterocyclic Compounds, Klingenberg (Ed.), part 3, chaps. IX-XII (1962) gives information on related compounds) in 10 ml concentrated sulfuric acid. The concentrated sulfuric acid should be added dropwise, with cooling (cf. U.S. Patent 3,429,889). Add 3 ml POCl$_3$ and stir at room temperature for 24 hours. Neutralize with NaHCO$_3$ to precipitate 2.3g (I). Filter; wash precipitate with

NaHCO$_3$ and recrystallize from acetonitrile. Dissolve 4.3 g (I) in 30 ml anisole and add 0.1 M methyl MgI in 50 ml anisole. Stir 12 hours and evaporate in vacuum or acidify with sulfuric acid, neutralize with NaHCO$_3$ and filter; wash to get 2.4 g N-benzyl analog of THC. For other N-analogs of unknown activity see JOC 33,2995(1968). Recover unreacted olivetol as usual.

The 5-aza analogs given in the JOC ref. seem to be active but they use the pyrone intermediate from certain routes of THC synthesis for a precursor. See U.S. Patent 3,493,579 (03 Feb 1970) for quinuclidine analogs and JOC 38,440(1973) for a different approach to N-analogs. See JOC 39,1546(1974) and HCA 56,519(1973) for other N-analogs.

Δ$^{1(6)}$ THC U.S. Patent 3,576,887

This synthetic route allows one to proceed from the alkylresorcinol dimethyl ether without using a compound of the verbenol or cyclohexanone type.

Synthesis of olivetol aldehyde (Aust. J. Chem. 21,2979 (1968)). To a stirred solution of phenyllithium (1.6g bromobenzene and 0.16g Li) in 50 ml ether, add 0.01M olivetol dimethyl ether (or analog -- see elsewhere here for preparation) in 5 ml ether and reflux 4 hours. Add 5 ml N-methylformanilide, reflux 1 hour and wash with 2X50 ml dilute sulfuric acid, 50 ml water, 25 ml saturated NaCl and dry, evaporate in vacuum the ether (can dissolve in benzene and filter through 100g of alumina) to get 60% yield of the dimethylolivetol aldehyde (I) (recrystallize from ether-pentane). Can recover unreacted starting material by refluxing the vacuum distillate 3 hours with excess 10% HCl, removing the organic layer and extracting the aqueous layer with ether: wash and dry, evaporate in vacuum the combined ether layers.

An alternative method for (I) (JACS 65,361(1943)). In a 200 ml 3 neck round bottom flask with a stirrer, a reflux condenser, a dropping funnel and a nitrogen inlet tube, introduce a rapid stream of nitrogen and in the stream issuing from the central neck, cut 1.5g of lithium into ca. 70 pieces and drop into the flask containing about 25 ml dry ether. Place the fittings in position, slow the nitrogen stream and add ¼ of the solution of 9.2g n-butyl-chloride in 25 ml dry ether. Start the stirring and add the rest of the n-butyl-chloride at a rate giving a gentle reflux. Continue stirring and reflux 2 hours and add 15 ml olivetol dimethyl ether in 25 ml dry ether. Reflux 2 hours and add dropwise a solution of 15 ml N-methylformanilide in 25 ml dry ether with stirring at a rate sufficient to produce refluxing. Continue stirring 1 hour, treat with 3% sulfuric acid and then pour into excess of this acid. Remove upper layer and extract aqueous layer twice with ether. Wash combined ether layers with dilute aqueous $NaHCO_3$ and water and dry, evaporate in vacuum the ether (can distill 148-52/0.3) to get 78% (I).

JACS 65,361(1943). A mixture of 6.5g (I) (or analog), 20 ml pyridine, 1 ml piperidine and 9g malonic acid is warmed on a steam bath 1 hour. Add another 1g malonic acid and heat another ½ hour. Reflux ½ hour and pour into excess iced 10% HCl, stirring occasionally over 2 hours. Filter and dry to get 6 g 2,6-dimethoxy-4-n-amycinnamic acid (II) (recrystallize from ethanol).

10g (II), 40 ml 80% isoprene and 40 ml dry xylene or toluene is heated in an autoclave at 185° C for 15 hours. Cool, dilute with 160 ml petroleum ether and shake with 100 ml saturated aq. Na_2CO_3. Let stand and separate the middle layer. Wash the middle layer with a mixture of petroleum ether and dilute aq. Na_2CO_3 and again separate the middle layer and treat with 75 ml 10% HCl and 75 ml ether. Shake, separate the aqueous layer and wash the ether 3 times with water. Dry and evaporate in vacuum the ether and dissolve the residue in petroleum ether. The solid which ppts. after about 10 minutes is unchanged (II). Filter and let stand in refrigerator overnight and dry and evaporate in vacuum to precipitate about 7 g of the 1-methyl-5 (2,6-dimethoxy-4-n-amylphenyl)-1-cyclohexene-4-COOH (III) (recrystallize from petroleum ether).

1g (III) in 5 ml dry ether is added to 10 ml 3M MeMgI (from 0.21g Mg and 1.2g methyl iodide) in ether, heated to 130° C to evaporate the solvent and the oil kept at a bath temperature of 165° C for ½ hour. Cool in dry ice-acetone bath and cautiously add ammonium chloride-ice water mix to decompose the excess Grignard reagent. Acidify with dilute HCl and extract with ether. Wash with NaCl, dilute K_2CO_3, NaCl and dry, evaporate in vacuum to get the dimethyl derivative (IV). Reflux (IV) in 25 ml benzene with 100 mg p-toluenesulfonic acid for 1 hour with a Dean-Stark trap and dry, evaporate in vacuum (or wash with $NaHCO_3$, NaCl first) to get the THC or analog.

Hydrolysis of benzopyrones (for synthesis see elsewhere here) will produce compounds of type (III) which will work in this synthesis. The hydrolysis proceeds as follows (JCS 926(1927)): Add 10g of the benzopyrone to 20g 30% NaOH, cool and shake 1 hour with 19 ml methylsulfate. Extract the oil with ether and dry, evaporate in vacuum to get the ester. Acidify the aqueous solution and filter, wash, dissolve ppt. in sodium carbonate and acidify, filter to get the free acid. Both the acid and the ester will work in this synthesis.

For a possible route to benzopyrones via condensation of isoprene and 3-CN-5-OH-7-alkyl-coumarin see JACS 82,5198(1960). See JMC 16,1200(1973) for another ref. on the pyrone route to THC.

Δ³ THC Analogs TET 23,77(1967)

11.6g 5-(1,2-dimethyl)-heptyl resorcinol or equimolar amount of olivetol or other analog, 9.2g 2-carbethoxy-5-methyl cyclohexanone (4-carbethoxy-1-methyl-3-cyclohexanone), 5 g $POCl_3$, 70 ml dry benzene (protect from moisture with $CaCl_2$ tube). Boil 5 minutes (HCl evolution) and let stand at room temperature 20 hours. Pour into 10% $NaHCO_3$, separate the benzene layer and wash with 3X50 ml 10% $NaHCO_3$. Dry and evaporate in vacuum the benzene and recrystallize from 50 ml ethyl acetate to get 6.6 g of the pyrone (I). 4.5g(I), 150 ml benzene; add dropwise to a solution prepared from 7.8 g Mg, 18 ml methyl iodide, and 90 ml ether. Reflux 20 hours and add 45 ml saturated NH_4Cl. Separate the organic layer and extract the aqueous phase with benzene. Combine the organic layer and benzene and dry, evaporate in vacuum to get the THC analog.

Δ³ THC analogs from Resorcinol TET 23,83(1967)

22g resorcinol, 36 g 4-carbethoxy-1-methyl-3-cyclohexanone, 20 g polyphosphoric acid; heat to 105° C and when the exothermic reaction which occurs subsides, heat at 140° C for one-half hour. Pour onto ice-water; filter; wash with water and recrystallize-ethanol to get 34 g of the pyrone (I). 6.4g (I), 8 ml caproyl-Cl or analog (for preparation see above reference, page 84); heat on oil bath (can use mineral oil) at 120° C until the exothermic reaction subsides (HCl evolution). Cool and pour into ethanol. Filter to get 8 g precipitate (II). 3.2g (II), 4.4g dry $AlCl_3$; heat on oil bath at 170° C for one hour. Cool and add HCl; filter and dissplve precipitate in 7 ml 2N NaOH. Filter and acidify with HCl to precipitate 1.4 g (III) (recrystallize-ethanol). Test this for activity. Use benzoyl-Cl or benzoic anhydride to esterify the OH group (this may not be necessary), methyl MgBr or methyl MgI to methylate the keto group, and sulfuric acid to dehydrate and hydrogenate as described elsewhere here to get the THC analog. Since the resulting THC analog has the side chain at the 6' position, it may not be active. This paper also gives a synthesis for THC analogs with the side chain in the 4' position, but again their activity in man is unknown. Verbenol, etc., should work in this synthesis, thus obviating the need for the methylation step.

$\Delta^{1(6)}$ THC JACS 88,367(1966)

1M olivetol or analog, 1M citral in 10% BF_3 etherate in benzene about eight hours at 5-10° C. Extract unreacted olivetol with dilute NaOH and evaporate in vacuum the ether to get about 20% yield of the trans THC, and 20% of the cis THC which can be converted to the active trans isomer by reacting with BBr_3 in methylene chloride at -20° C for 1½ hours. (TL 4947(1969)). Alternatively, the reaction can be carried out in 1% BF_3 etherate in methylene chloride to get 20% Δ^1 THC.

Δ^3 THC Analogs JACS 63,1971(1941) cf. CA 82,170672-3 (1975)

7.6g 5-n-heptyl resorcinol or equimolar amount analog, 6.6g (0.037M) 4-carbethoxy-1-methyl-3-cyclohexanone or analog, 5.8g $POCl_3$ in 60 ml benzene. Reflux 5 hours, cool and pour into $NaHCO_3$ to get about 6 g THC analog and 1 g more by concentrating the mother liquor, or proceed as described elsewhere here to recover unreacted resorcinol. 3-carbethoxy-1-methyl-2 or 4-cyclohexanone, 2-carbethoxy-cyclohexanone, etc. will probably also give active THC analogs.

Δ^3 THC Analogs JCS 952(1949)

1.75 g 2-Br-4-methyl-benzoic acid, 1.5 g olivetol or analog, 10 ml 1N Na OH and heat to boiling; add 0.5ml $CuSO_4$. Filter; wash with ethanol and recrystallize from ethanol to get (I). 10 g (I) in 150 ml benzene; add to methyl-MgI prepared from 47.5 g methyl iodide, 8 g Mg, 120 ml ether. Reflux fifteen hours, cool and pour on ice. Add saturated NH_4Cl and separate the ether. Wash two times with water and dry and evaporate in vacuum the ether to get the THC.

Precursors For THC Synthesis

p-Menthatriene (1,5,8) BER 89,2493(1956)

90g d(+) carvone (ℓ (-) carvone or racemic carvone probably will work also) in 150 ml ether; add dropwise with stirring to 7.5 g lithium aluminum hydride in ether. Heat one hour on water bath; cool and carefully add water and then ice cold dilute sulfuric acid. Separate the ether and extract the aqueous layer with ether; dry and evaporate in vacuum the combined ether to get about 60 g product (can distill 65/14).

(+) Cis and Trans p-Menthadien-(2,8)-OL-1

Method 1: LAC 674,93(1964) cf. BSC 3961(1971), JOC 38,1684(1973)

136g(+) limonene in 2 liters methanol; 2g bengal rose dye. Illuminate with a high voltage Hg lamp (e.g., HgH 5000) for fourteen hours or until about 1M of O_2 is taken up. Evaporate the methanol at 0-10° C to about 500 ml and then stir with ice cooling and add this solution dropwise to solution of 250 g Na_2SO_3 in 1.5 liters water and continue stirring for twelve hours. Heat two hours at 70° C and extract with ether and dry, evaporate in vacuum (can distill with addition of Na_2CO_3 at 40-70/0.2) to get about 100 g mixture containing about 40% product which can be purified by fractional distillation.

Method 2: HCA 48,1665(1965)

Convert (+) Δ^3 carene to (+) trans-4-acetoxycarane (I) via (+) trans-4-OH-carane. Reflux 50 g (I) for 45 minutes (180° C oil bath under N_2 or Argon). Cool and can distill (57/10) to get about 25 g mixture of Δ^3 and Δ^4 carene (residue is unchanged starting material) containing about 60% Δ^4 isomer.

Alternatively, to 150 ml ethylene diamine add portionwise with stirring at 110° C under Argon or N_2, 5.3 g Li metal; after one hour add dropwise 110 g (+) Δ^3 carene. After one hour cool to 4° C and add water. Extract with ether, wash the ether five times with water and dry, evaporate in vacuum to get 100 g of a mix containing about 40% (+) Δ^4 carene (can separate by fractional distillation).

Δ^4 carene can also be obtained from Δ^3 carene as follows: (JCS (C) 46(1966)): Dissolve 1 g Δ^3 carene in 50 ml propionic acid and heat at a suitable temperature (e.g., one-half hour at room temperature may do) in presence of ½g Palladium-Carbon catalyst (5%) in ethanol and filter, evaporate in vacuum (can distill 63.5/19.5). See J.Soc. Cosmet. Chem. 22,249(1971) for a review of (+) Δ^3 carene chemistry.

Δ^2 Carene oxide (2-epoxycarene) LAC 687,22(1965), (cf. TL 2335(1966), and CA 68:22063(1968))

To 136 g Δ^4 carene in 330 ml methylene chloride and 120 g anhydrous sodium acetate, add dropwise with vigorous stirring in an ice bath, 167 g of 50% peracetic acid and continue stirring for ten hours. Heat to boiling for two hours, cool, wash with water, sodium carbonate, water, and dry, evaporate in vacuum the methylene chloride to get about 100 g p-menthadieneol. Apparently (CA

68,22063(1968)) substituting sodium carbonate for sodium acetate results in the production of Δ^2 carene oxide (2-epoxycarene) in about 50% yield (can distill 63/7).

4-Carbethoxy-1-methyl-cyclohexanone LAC 630,78(1960)

Cool 20 g of sodium metal in 325 ml ethanol to -15° C in an ice-salt bath and add in small amounts over one hour a solution of 100 g 3-methyl-cyclohexanone and 150 g diethyloxalate (keep temperature below -10° C). Keep three hours in cold and then twelve hours at room temperature. Add solution of 1.3 L of water, 60 ml 2N sulfuric acid. Separate the yellow-brown oil and extract the water with ether or $CHCl_3$ until the yellow is removed. Combine the oil and the extract and distill the solvent and the unchanged starting material (100° C bath, 13 mm). Slowly heat the residue in a one-half liter flask with air cooling. CO_2 evolution starts at 160° C. Continue heating to 220° C and keep at this temperature for 1½ hours or until a test with 1% ethanol-$FeCl_3$ solution shows the end of the reaction by a violet color (unconverted material gives a brown color). Can distill two times in Vigreux column to give about 83 g of oily colorless product.

(-) Verbenol JCS 2232(1961)

Racemic alpha-pinene will yield racemic verbenol which will give one-half the yield of (-) verbenol.

27 g (-) alpha-pinene in 500 ml dry benzene; heat and keep temperature at 60-65° C throughout. Add with stirring over 20 minutes 84 g dry (dry over P_2O_5) lead tetra-acetate. Stir one-half hour; cool and filter and add filtrate to water. Filter and evaporate in vacuum the benzene layer (can distill 96-7/9) to get 21.2 g cis-2-acetoxy-pin-3-ene(I). 5 g (I) in 25 ml glacial acetic acid; keep at 20° C for one-half hour and add water and extract with ether. Wash the extract with aqueous Na_2CO_3 and evaporate in vacuum the ether (can distill 97-8/9) to get 4.3 g trans verbenyl acetate (II). Hydrolyze (II) with NaOH to give the (-) cis and trans verbenol. For other methods of producing verbenol see CA 37,361(1943), CA 57,16772(1962) and BSC 2184(1964), JCS (B) 1259(1967). The last paper also gives a method for converting (-) beta-pinene to (-) alpha-pinene. See also CA 65,2312(66).

5-Alkyl Resorcinols from Acyl Resorcinols CA 72,66922(1970)

Compounds I-III may be able to give active THC analogs if used in place of olivetol for synthesis.

45 g 1-(3,5-dimethoxyphenyl)-1-hexanone(I) or analog (for

preparation see the following methods) in 400 ml ether and 0.3 M methyl-MgI in 150 ml ether react to give 49 g 2-(3,5-dimethoxyphenyl)-2-heptanol(II). Heat 49 g (II) with 1 ml 20% sulfuric acid to 105-125° C/30mm for 1½ hours to get 34 g of the 2-heptene compound (III). 33 g (III) in 100 ml ethanol, 6 g Raney-Ni, 1500 PSI hydrogen, 150° C to get 26 g of the 2-heptane (IV). 26 g (IV), 118 ml 57% hydrogen iodide; add 156 ml acetic anhydride and heat at 155° C for two hours to get 22 g of the resorcinol.

5-Alkyl Resorcinals BER 69,1644(1936)

25 g ethyl-3,4,5-trimethoxybenzoyl acetate and 2.1 g Na in 100 ml ethanol; warm to react. Add 2 g n-propyl iodide (or n-amyl iodide, etc.) and heat twelve hours on steam bath; neutralize and distill off the ethanol. Extract with ether and dry, evaporate in vacuum to get about 32 g of the alkyl acetate (I). Heat 22 g (I) in 5% KOH in ethanol for one hour at 50° C to get 14 g 3,4,5-trimethoxyvalerophenone (II), which crystallizes on standing. 11 g (II), 600 ml ethanol, 60 g Na; warm and after Na is dissolved, add 2 L water. Acidify with HCl, distill off the ethanol and extract with ether. Dry, evaporate in vacuum the ether to get 7.8 g olivetol dimethyl ether (or analog) (III). 7.2 g (III), 70 ml hydrogen iodide; boil two hours and distill (164/760) to get olivetol.

Olivetol HCA 52,1132(1969)

Reduce 3,5-dimethoxybenzoic acid with lithium aluminum hydride to 3,5-dimethoxybenzyl alcohol (I). to 10.5 g (I) in 100 ml methylene chloride at 0° C add 15 g PBr_3; warm to room temperature and stir for one hour. Add a little ice water and then more methylene chloride. Separate and then dry, evaporate in vacuum the methylene chloride. Add petroleum ether to precipitate about 11.5 g of the benzyl bromide (II). To 9.25 g (II), 15 g CuI, 800 ml ether at 0° C, add butyl (or other alkyl)-Li (16% in hexane), and stir for four hours at 0° C. Add saturated NH_4Cl and extract with ether. Dry and evaporate in vacuum the ether (can distill 100/0.001) to get about 4.5 g olivetol dimethyl ether (III) or analog. Distill water from a mixture of 90 ml pyridine, 100 ml concentrated HCl until temperature is 210° C. Cool to 140 ° C and add 4.4 g (III); reflux two hours under N_2. Cool and pour into water. Extract with ether and wash with $NaHCO_3$. Make pH 7 and dry, evaporate in vacuum to get 3.8 g olivetol which can be chromatographed on 200 g silica gel (elute with $CHCl_3$) or distill (130/0.001) to purify.

5-Alkyl Resorcinols TET 23,77(1967)

Since the method as given originally leads to 4-alkyl resorcinols which do not produce an active THC, it is here modified to give the 5-alkyl isomers. The method is illustrated for 1.2-dimethyl-heptyl resorcinol which gives a much more active THC than olivetol.

Convert 3,5-dihydroxyacetophenone (5-acetyl resorcinol) to 3,5-dimethoxyacetophenone(I) in the usual way with dimethylsulfate.

To 24 g Mg, 1 crystal I_2, 100 ml ether, add dropwise under N_2, 180 g 2-Br-heptane in 100 ml ether over one hour and then reflux two hours. Add over 1½ hours a solution of 90 g (I) in 200 ml tetrahydrofuran and reflux 10 hours. Cool and add 180 ml saturated NH_4Cl; decant the solvents and extract the residue with tetrahydrofuran. Combine the solvents and the tetrahydrofuran and dry, evaporate in vacuum. Add a few drops 20% sulfuric acid to the residual oil and evaporate in vacuum the water (oil bath temperature 120-130° C/10mm). Distill the oil at oil bath temperature 285° C/0.2. Fraction boiling 128-140/0.2 yields about 60 g 2-(3,5-dimethoxyphenyl)-3-methyl-2-octene(II). If saponified and used to synthesize a THC, this might give an active product, thus disposing of the necessity of the next step. Hydrogenate 50 g (II) in 100 ml ethanol, 2-3 atmospheres H_2, 0.6 g 10% Palladium-Carbon catalyst for two hours, or until no more H_2 uptake (or use the $NaBH_4$-Ni method described at the start). Filter and dry, evaporate in vacuum, and distill the residual oil (110-17/0.1) to get 42 g of the octane (III). 40 g (III), 100 ml 48% HBr, 320 ml glacial acetic acid and reflux four hours. Pour on ice and take pH to 4.5 with 10 N NaOH and extract with ether. Extract the ether with 3X150 ml 2N NaOH; acidify the combined NaOH extracts with glacial acetic acid and extract with ether. Dry and evaporate in vacuum, and distill the oil (159/0.1) to get 20 g 5-1, 2-dimethylheptyl resorcinol.

5-Alkyl Resorcinols JACS 61,232(1939)

Convert benzoic acid to 3,5-dihydroxybenzoic acid (alpha-resorcyclic acid) (I). 50 g (I), 134 g dimethylsulfate, 60 g NaOH, 300 ml water; add 35 g NaOH and reflux to obtain about 50 g 3,5-dimethoxybenzoic acid (II) which is converted to dimethoxybenzoyl chloride (III) with PCl_5. Extract the (III) with ether and filter. Saturate the ether with NH_3 at 0° C and filter. Wash with ether and water and recrystallize from hot water to get 3,5-dimethoxybenzamide (IV). To a solution of 1 M of n-hexyl bromide (or 1,2-dimethylheptyl bromide, etc.) add 24.3 g Mg in 200 ml ether to prepare the Grignard reagent. Then rapidly add 46 g (IV); add 300 ml ether and reflux and stir two days, excluding moisture and air. Add a mixture of ice and water and 80 ml concentrated sulfuric acid. Separate and dry, evaporate in vacuum the ether layer to get about 50 g of the dimethoxyalkyl benzyl ketone (V). Recrystallize from dilute ethanol. Add 0.2 M (V), 20.8 g 100% hydrazine hydrate, 75 ml ethanol; reflux six hours. Evaporate in vacuum and heat the residual oil with 82 g powdered KOH in oil bath about 225⁰ C until N_2 evolution ceases. Can distill or recrystallize from 95% ethanol to get the dimethoxyalkyl benzene (VI). 0.025 M (VI), 40 ml glacial acetic acid, 15 ml 48% HBr; reflux four hours and pour into ice water; decolorize with a little Na bisulfite, neutralize with $NaHCO_3$ and extract with ether. Wash the extract with 10% NaOH and separate and acidify the basic solution. Extract with ether and dry, evaporate in vacuum the extract to get the 5-alkyl resorcinol. Distill or recrystallize from water to purify. Dry, evaporate in vacuum the first ether extract to recover starting material.

Olivetol LAC 630,77(1960), JCS 311(1945), cf. JACS 89,6734 (1967)

Dissolve 100 g malonic acid in 360 g dry pyridine and heat 48-52° C for forty hours with 100 g n-hexaldehyde (n-capronaldehyde) or homolog. Cool in ice bath and with good stirring add dropwise 150 ml ice cold concentrated sulfuric acid (keep temperature below 5° C). After addition add water to dissolve the precipitate and extract with ether two times. Dry, evaporate in vacuum the ether and distill (70/0.7 or 102/5) to get about 98 g 2-octenoic acid (I). 95 g (I) in 300 ml ether; cool to -5° C and slowly add a solution of an excess of diazomethane in ether dried over KOH and let react for about one hour. Let stand twelve hours, evaporate in vacuum and distill (91/17) to get about 94 g clear methyl-2-octenoate (II).

To 16.3 g Na in 210 ml ethanol add 93 g ethyl-acetoacetate (ethyl-3-oxo-butanoate), heat to boil and add dropwise 92 g (II) over 20 minutes. Stir and reflux five hours and cool to precipitate. Filter, wash with ethanol and dissolve precipitate in 800 ml water. Cool to 0° C and slowly add 80 ml ice cold concentrated HCl to precipitate. Filter, wash with water and ligroin to get about 108 g 6-carbethoxy-4,5-dihydro-olivetol (III) (recrystallize from petroleum ether). To 104 g (III) in 260 ml glacial acetic acid at room temperature with good stirring, add dropwise over one hour 69 ml Bromine. Heat four to five hours at 60° C, cool and add 300 ml water and let stand twelve hours. Oil separates which will precipitate on agitation and rubbing. Filter, wash with water until colorless (recrystallize from ligroin, recrystallize from glacial acetic acid and precipitate with water) to get about 86 g 6-carbethoxy-2,4-dibromo-olivetol (IV). 0.035 g Palladium-Carbon catalyst in 25 ml hydrogenation bottle. Saturate with H_2 (pressure - 2.8 Kg/cm^2) and add 0.33g (IV) in 5 ml glacial acetic acid, which takes up 39.5 cm^3 H_2 at atmospheric pressure over 1½ hours at 60-70° C. Filter and acidify at 0° C with ice cold 6N HCl. Extract with ether and dry, evaporate in vacuum. Recrystallize the oil from ligroin and then from glacial acetic acid by adding water to get about 0.2 g 6-carbethoxyolivetol (V). (IV) can also be hydrogenated at room temperature and atmospheric pressure over ½g Palladium-Carbon catalyst by dissolving 70 g in 500 ml 1N NaOH. Heat 35 g (V) with 45 g NaOH in 170 ml water for two hours or until no more CO_2 is evolved. Cool, acidify with 6N HCl and boil 3 minutes. Extract the oil with ether and dry, evaporate in vacuum the ether (can distill on Vigreux column 123/0.01, oil bath 160° C) and let oil stand in refrigerator until crystalline to get about 21 g olivetol. See CA 70,77495t(1969) for another variant of this procedure.

1',1'-Dimethylolivetol HCA 52,1127(1969)

Prepare 3,5-dimethoxy benzyl alcohol by reducing the acid with lithium aluminum hydride as described elsewhere here, by hydrogenating the aldehyde (**2-3 atmospheres H$_2$, room temperature, PtO$_2$** in ethanol -- or by the NaBH$_4$ method), in five steps as described in JACS 70,666(1948), or prepare (II) directly by the doborane procedure.

Add with stirring 22.5 g SOCl$_2$ in 100 ml ether in 20 ml portions to a solution of 15 g 3,5-dimethoxybenzyl alcohol, 1 ml pyridine and 200 ml ether. Let stand and wash with 2X100 ml cold water; separate and dry, evaporate in vacuum the ether to get 16 g 3,5-dimethoxybenzyl chloride (I). Recrystallize from petroleum ether. 16 g (I), 300 ml ethanol, 30 g NaCN, 75 ml water; reflux three hours and pour onto 400 g ice. After ice melts, filter and recrystallize precipitate from petroleum ether to get about 14 g 3,5-dimethoxybenzyl CN (II). 5 g 50% NaH in mineral oil; wash three times with pentane or hexane; fill flask with N$_2$ or argon and add dimethoxyethane or dimethylformamide (freshly distilled from K if possible). Stir and add 9 ml methyl iodide. Carefully add 8 g (II) and stir twelve hours. Add ice water and neutralize with NaHCO$_3$ to pH 7-8. Extract with ether and dry, evaporate in vacuum the ether (can distill 170/0.1) to get about 9 g alpha, alpha-dimethyl-3,5-

dimethoxyphenylacetonitrile (III). Add 1.5 g (III) to a solution of 0.45 g Mg, 2.5 g n-propyl Br (freshly distilled if possible) in 30 ml ether. Reflux sixty-five hours. Add 2N sulfuric acid and heat two hours on water bath. Cool and extract with ether and dry, evaporate in vacuum (can distill 135/0.001) to get about 1.75 g 2-methyl-2-(3,5-dimethoxyphenyl) hexanone-3 (IV). Dissolve 4 g (IV) in 50 g ethane dithiol and saturate at 0° C with HCl gas (take care to exclude water). Stir the solution in a sealed container forty-eight hours at room temperature and then basify with $NaHCO_3$. Extract with ether and dry, evaporate in vacuum (or dry and evaporate in vacuum two hours at 70/12 and distill at 130/0.001) to get about 5 g of the thioketal (V). Reflux 5.3 g (V), 100 g Raney-Ni, 2 L ethanol (or use $NaBH_4$ procedure) for thirty hours. Cool and filter (Celite), evaporate in vacuum and distill residue (115/0.001) to get 3.7 g of the hexane which is saponified as described for the dimethyl ether of olivetol above to give about 2.5 g of the title compound (can distill 150/0.001).

Olivetol ACS 24,716(1970)

Prepare 3,5-dimethoxybenzoic acid as described elsewhere here, and to a solution of 18.2 g in 250 ml dry tetrahydrofuran under N^2, add 1 g 85% LiH, stir for fourteen hours and then reflux for one-half hour. Add a solution of about 1.3 M butyllithium in ether (Org. Rxns. 6,352(1957)) with stirring and ice cooling until the reaction mixture gives a positive Gilman test (JACS 47,2002(1925)). Then add 500 ml ice water, extract with ether and dry, evaporate in vacuum the organic phase to get a yellow oil which is dissolved in an equal amount of absolute ethanol; left in refrigerator twelve hours to precipitate. Filter and evaporate in vacuum the ethanol to one-half volume to give more precipitate for a total of 18 g 1-(3,5-dimethoxyphenyl)-1 pentanone (I). 5.64 g (I) in 200 ml methanol; 0.66 g 20% $Pd(OH)_2$ on carbon (TL 1663(1967)) and hydrogenate at room temperature and atmospheric pressure over two to three hours (or use other reducing method as described here). Filter and evaporate in vacuum to get olivetol dimethyl ether (II). 4.88 g (II), 40 ml HI (density 1.7, decolorized with red phosphorous) and stir three hours at 115-125° C under N_2. Dry, evaporate in vacuum or pour into 100 ml ice water and extract with methylene chloride; wash methylene chloride with water and dry, evaporate in vacuum (can distill 160-170/3-4) to get 3.5 g olivetol.

5-Alkyl Resorcinols JOC 33,687(1968), JACS 71,1624,1628 (1949)

Illustrated for 1,2-dimethylheptyl compound.

110 g powdered 3,5-dimethoxybenzamide (preparation given elsewhere here), five times excess of methylMgI and reflux sixteen hours. Add 1.2 L concentrated HCl and 1200 g ice and let stand sixteen hours with occasional shaking. Extract with ether, dry, evaporate in vacuum and distill (115-128/0.3). Let stand in refrigerator to precipitate. Wash precipitate with petroleum ether and recrystallize from petroleum ether to get about 60 g 3,5-dimethoxyacetophenone (I). 83 g (I) in 50 ml methanol. Add dropwise with stirring to solution of 18 g NaBH$_4$ in 300 ml methanol and 1 g NaOH. Reflux 30 minutes and concentrate by distilling. Add about 100 ml water during distillation. Evaporate the methanol, cool and extract with ether. Dry, filter, concentrate, and distill (124/0.65) to get about 80 g 3,5-dimethoxyphenyl ethanol (II). 18 ml PBr$_3$ in 70 ml ether; add dropwise with stirring over one hour to 28.5 g (II) in 70 ml ether cooled in an ice bath. Warm to room temperature; reflux two hours on steam bath; cool and pour into 200 g ice. Shake and extract with ether 3 times; wash ether with 10% NaHCO$_3$ and water and dry, filter. Concentrate on steam bath and then add with stirring under anhydrous conditions to 42 g diethyl-n-amylmalonate in 300 ml ethanol in which has been dissolved 4.6 g Na metal. Stir 1½ hours at room temperature and then heat to distill off the ether and complete the reaction. When the distillation head temperature reaches 78° C add water and continue distilling until temperature reaches 99° C. Cool and extract with 3X250 ml ether and evaporate in vacuum the ether. Dissolve residue in 180 ml ethylene glycol and 35 g NaOH by stirring six hours at 160° C. Cool and add 1500 ml water and wash with ether. Acidify the aqueous phase and extract with 4X200 ml ether. Evaporate in vacuum the ether or evaporate on steam bath and dissolve the residue in 150 ml xylene. Evaporate residual ether and water until head temperature reaches 140° C and reflux six hours. Evaporate in vacuum to get about 30 g oily alpha-amyl-beta-methyl-hydrocinnamic acid (III). 14.5 g (III), 5 g lithium aluminum hydride, 250 ml ether; reflux 6 hours. Cool and carefully add methanol, water and dilute HCl. Separate the aqueous layer, saturate with NaCl and extract with ether. Wash ether with NaHCO$_3$ (acidify NaHCO$_3$ extract to precipitate starting material) and dry, evaporate in vacuum the ether to get about 12 g 2-amyl-3(3,5-dimethoxyphenyl)-butanol (IV). 11.7 g (IV) and 8.5 g p-toluenesulfonyl Cl each dissolved in 20 ml pyridine. Cool in ice bath and combine. Place in freezer about sixteen hours, pour over ice and extract with ether.

Wash ether with cool, dilute HCl until HCl extract is acidic. The combined HCl extracts are then acidified and extracted with ether. Wash the ether with $NaHCO_3$, dry and then add dropwise with stirring to 3 g lithium aluminum hydride in 75 ml ether. Reflux 4½ hours and work up as for (IV). (Can chromatograph the undistilled product on activated alumina and elute with 600 ml petroleum ether, then 200 ml methanol; concentrate and distill (94/0.001) the petroleum ether to get product; concentrate the methanol to give starting material). Yield is about 8 g 2-(3,5-dimethoxyphenyl)-3-methyl-octane (V). Convert (V) to the title compound by refluxing in 48% HBr in glacial acetic acid as described elsewhere here.

An alternative route from (II) to (V) involves adding (II) and diethylmethylmalonate to prepare dimethyl-3,5-dimethoxyhydrocinnamic acid as described for the preparation of (III). Then dehydrate and hydrolyze to dimethyl-3,5-dimethoxycinnamic acid which is hydrogenated to the alcohol and converted in several steps to (V).

5-Alkylresorcinols Aust. J. Chem. 21,2979(1968)

Reflux 6.9 g triphenylphosphine and 6.6 g lauryl bromide (or equimolar amount of homolog) in 40 ml xylene for 60 hours. Remove solvent and wash residue with 5X20 ml ether (by decanting) to get 11 g lauryl triphenylphosphonium bromide (I). To a stirred suspension of 5.6 g (0.011M)(I) in 50 ml ether add 0.01M butyllithium solution (see Organic Reactions 8,258(1954) for preparation). Stir ½ hour at room temperature and slowly add 1.66 g 3,5-dimethoxybenzaldehyde (preparation given elsewhere here) in 10 ml ether over ½ hour. After 15 hours, filter, wash filtrate with water and dry, evaporate in vacuum. Dilute residue with pentane, filter and remove solvent. Dissolve the residual oil in 25 ml ethyl acetate and hydrogenate over 0.1g Adams catalyst at one atmosphere and room temperature for 2 hours. Filter and evaporate in vacuum to get the 5-alkylresorcinol dimethyl ether which can be recrystallized from pentane and demethylated as described elsewhere here.

5-Alkylresorcinols Aust. J. Chem. 21,2979(1968)

Add 2.17 g 3,5-dimethoxybenzoyl chloride (see BER 41,1329 (1980) or elsewhere here for preparation) in 2.5 ml ether over 10 minutes to a stirred solution of 0.42g diazomethane and 1.01 g triethylamine in ether at -5° C. Keep 10 hours at 0° C, filter, wash precipitate with 20 ml ether and evaporate in vacuum the combined filtrates to get 1.9g diazo-3,5-dimethoxyacetophenone(I). Recrystallize from benzene-cyclohexane. To 1.5g(I) in 15 ml ethanol add 1.23g pyridinium perchlorate in 2 ml pyridine. Reflux 2 hours, cool and add 5 ml water. Filter to get 1.5g 3,5-dimethoxyphenylacyl pyridinium perchlorate (II). Recrystallize from ethanol. To a stirred suspension of 0.13 g Na hydride in 10 ml dry, acid free dimethylacetamide at 10° C under nitrogen, add 1.79g (II) in 10 ml dimethylacetamide and shake for 15 minutes. Add 0.0075M propyl iodide or homolog and keep 12 hours at room temperature. Heat 2 hours at 90° C and cool to room temperature. Add 3g zinc dust and 5 ml glacial acetic acid; stir at room temperature four hours and filter. Add 20 ml water to filtrate and extract with 50 ml ether. Wash the ether layer with 2X50 ml 10% K_2CO_3, 50 ml water and 20 ml saturated NaCl. Dry and evaporate in vacuum to get oily 3,5-dimethoxyphenylbutyl ketone (can chromatograph on 50 g alumina and elute with 3:1 petroleum ether:ether). Recrystallize from ethanol. Hydrogenolysis (see Aust. J. Chem. 18,2015(1965) or elsewhere here) gives the olivetol dimethyl ether.

5-Alkylresorcinols Aust. J. Chem. 24,2655(1971)

The method is illustrated for olivetol preparation, but substituted thiophens can be used to get olivetol homologs.

To a stirred solution of 45g 3,5-dimethoxybenzoyl chloride and 17.4g thiophen in 300 ml benzene at 0° C, add dropwise 10.5g freshly distilled stannic chloride. Stir one hour at room temperature and add 200 ml 3% aqueous HCl. Separate the benzene layer and wash the aqueous layer with benzene. Dry and evaporate in vacuum the combined benzene layers and distill the red residue (250° C bath/4.5) to get 45g 2-(3,5-dimethoxybenzoyl) thiophen(I). Recrystallize from petroleum ether. Add a solution of 21g $AlCl_3$ in 160 ml ether to a stirred suspension of 6.1g lithium aluminum hydride in 140 ml ether. After 5 minutes add a solution of 39g(I) in 300 ml ether at a rate giving a gentle reflux. Reflux and stir 1 hour; cool in an ice bath and treat dropwise with 50 ml water, then 50 ml 6N aqueous sulfuric acid. Separate the layers, extract the aqueous layer with 3X100 ml ether and dry, evaporate in vacuum the combined ether layers. Can distill the residue (230° C bath/5mm) to get 27g oily 2-(3,5-dimethoxybenzyl) thiophen (II). Recrystallize from petroleum ether. Reflux a solution of 5g (II) in 700 ml ethanol with W-7 Raney Nickel prepared from Ni-Al alloy (see Org. Synthesis Coll. Vol III,176(1955)) for 6 hours. Filter, evaporate in vacuum and can distill (140/0.01) to get about 2.2g oily olivetol dimethyl ether which can be reduced to olivetol as described elsewhere here.

The use of the novel reduction methods described at the beginning of this section would render this method much simpler.

5-Alkylresorcinols German Patent 2,002,815 (30 July 1970)

In a 2 liter, 3 necked flask with a stirrer, dropping funnel, thermometer, reflux head, nitrogen stream and mercury manometer (if available) stir 230 ml dry methanol and 32.4g sodium methoxide under nitrogen until dissolved. Add 110g diethylmalonate and stir 10 minutes. Add portionwise 75g 90% pure 3-nonene-2-one (for olivetol-preparation below) keeping the temperature below the boiling point (50-60° C). Stir and reflux 3 hours, then cool to room temperature, neutralize with about 50 ml concentrated HCl and let stand overnight. Evaporate in a vacuum and dissolve the residue in 200 ml 1N HCl and 800 ml ethylacetate. Separate and wash the ethylacetate with 2X300 ml water and extract with saturated $NaHCO_3$ until a small portion gives no turbidity upon acidification (about 5X200 ml). Carefully acidify the combined $NaHCO_3$ extracts and then extract with 3X300 ml ether. Dry and evaporate in vacuum the ether (can dry under vacuum several days) to get 6-n-pentyl-2-OH-4-oxo-cyclohex-2-ene-1-methyl-carboxylate(I). 4.8 g (I) and 100 ml glacial acetic acid are stirred vigorously at 75° C until dissolved. Cool and keep temperature between 5 and 10° C while adding a solution of 3.9g bromine in 10 ml glacial acetic acid dropwise over 1 hour. Stir 1 hour at room temperature then 3 hours on a steam bath. Evaporate in vacuum and dissolve the residual oil in 200 ml ether. Wash with 2X25 ml 10% sodium dithionite, 2X25 ml saturated $NaHCO_3$ and water and dry, evaporate in vacuum to get olivetol (or analog) (can distill at 125-130/0.05).

Alternatively, to 4.8g (I) ad 5.6g iodine in 200 ml glacial acetic acid. Stir and reflux 10 hours, evaporate in vacuum, dissolve residue in 250 ml ether and proceed as above to get olivetol.

A third alternative is to stir 12.2g (I) in 100 ml glacial acetic acid at 25° C with vigorous stirring until well suspended. Cool and keep temperature at 5-10° C while adding dropwise 22.4g cupric bromide dissolved in 25 ml glacial acetic acid over 1 hour. Stir 1 hour at room temperature then 3 hours on a steam bath and evaporate in vacuum. Dissolve the residue in 200 ml water and 300 ml ether. Wash the ether layer with 2X50 ml 10% sodium dithionite, 2X35 ml saturated $NaHCO_3$ and 75 ml water and dry, evaporate in vacuum to get olivetol (or analog).

For the 1,2-dimethylheptyl homolog proceed as follows. Combine

15g 5,6-dimethylundec-3-ene-2-one with 19g diethylmalonate as described above to get (I). Brominate 20.2g (I) with 12g bromine over 2 hours as described and stir 1 hour at room temperature. Add 500 ml water and let stand overnight at 5-10° C. Filter, wash precipitate with about 4X75 ml cold water and dry in vacuum at 50° C to get 26g 3-bromo-2-OH-4-oxo-6(l,2-dimethylheptyl)-cyclohex-2-ene-1-methylcarboxylate(II). In a 3 liter 3 necked flask with a stirrer, thermometer, reflux head and Dean-Stark trap, add 350g (II) and 522g pyridine hydrochloride and heat on oil bath at 90° C 4 hours. Heat with the heating mantle (removing volatiles with the Dean-Stark trap) until the internal temperature reaches 190-200° C and hold at this temperature 2 hours. Cool to room temperature and shake with 3 liters ether and 660 ml 1.2N HCl and then 2X300 ml water. Extract the ether solution with 4X350 ml 10% NaOH and then extract the combined NaOH extracts with 2X300 ml ether. Acidify the alkaline solution with about 700 ml concentrated HCl and extract with 3X800 ml ether. Wash the combined ether extracts with 3X300 ml 10% sodium dithionite, 2X300 ml saturated NaHCO$_3$ and 300 ml water and dry, evaporate in vacuum the ether to get the 5-(l,2-dimethyl-heptyl) resorcinol.

To prepare the 3-nonene-2-one condense excess acetone with n-hexaldehyde (or 2,3-dimethyloctanal for 5,6-dimethyl-undec-3-ene-2-one) in the presence of NaOH in an inert medium if desired (benzene, toluene, xylene, etc.), at 10-70° C to get (I). Dehydrate (I) with sodium sulfate or cupric sulfate in an inert medium at reflux temperature or simply reflux in benzene, xylene or toluene.

5-Alkylresorcinols JOC 37,2901(1972)

For 5-alkylresorcinols see Chem. and Ind. 685(1973) also.

This is an improved version of a previously given synthesis (LAC 630,71(1960)). The ethanol used is distilled from Ca ethoxide; dimethoxyethane from potassium. Cupric bromide is produced from cupric oxide and 5% excess of HBr, plus sufficient bromine to remove the milkiness on addition of a drop of the mixture to water; concentrate and dry, evaporate in vacuum over KOH flakes.

650g (5.3M) ethyl chloroacetate and 880g (5.3M) triethyl phosphite are mixed and placed in a 3 liter flask fitted with a thermometer and condenser under nitrogen. Heat and stir and slowly bring to 125° C. Discontinue heating as ethyl chloride evolution proceeds over ½ hour. Heat to 160° C over a 75 minute period and keep at 160° C 8 hours. Cool, distill (e.g., through 12″

Vigreux column) (74-7/0.03) to get 96% yield of triethyl-phosphonoacetate (I). In a 3 liter flask fitted with a stirrer, dropping funnel and condenser, place 45.3g NaH (1M in mineral oil) and 1 liter of dry ether. Flush with nitrogen and keep at positive nitrogen pressure. Stir in ice bath while 224 g (1M) (I) is added dropwise over 75 min. Stir and reflux 1 hour (H_2 evolution stops). Cool in ice-salt bath and add 1M of aldehyde (e.g., hexaldehyde for olivetol) over 1 hour. Continue to cool and stir an additional 10 minutes and then slowly bring to reflux and reflux for 10 minutes (ppt. prevents stirring). Decant the ether and dissolve the oil layer in 500 ml warm water and separate the upper organic layer. Extract the aqueous layer with 200 ml ether and extract the combined organic solutions with 200 ml saturated $NaHCO_3$. Dry and evaporate in vacuum (can distill) to get the ethyl-β-alkylacrylates in about 90% yield (II).

In a flask with nitrogen and fittings as in preceding step, add 156 g ethyl acetoacetate to Na ethoxide from 25.3g Na and 500 ml dry ethanol, and stir and reflux ½ hour. Add 1M of (II) dropwise over 90 minutes and reflux 20 hours. Cool in ice, filter, wash ppt. with 500 ml ice cold absolute ethanol and several times with portions of ether and dry, evaporate in vacuum to get the dione Na salt (III) in about 80% yield (for olivetol precursor). In a 250 ml flask place 0.1M (III), 100 ml 1,2-dimethoxyethane and flush with nitrogen and stir at room temperature while 45g of cupric bromide is added portionwise over 5 minutes, under a nitrogen stream. Stir ½ hour and then reflux and stir 1 hour. Cool and evaporate (keep temperature below 50° C) but do not remove more than about 65 ml dimethoxyethane. Dilute the remaining solution with 200 ml benzene and filter. Wash the ppt. with 50 ml benzene and evaporate the combined benzene filtrates (keep temperature below 50° C). Dissolve the bromodione in 100 ml dimethylformamide and put in 500 ml flask under nitrogen. Stir and heat slowly until reflux and then heat to 150° C and reflux 4 hours. Cool, pour into 500 ml water, extract with 3X100 ml dichloromethane and dry, evaporate in vacuum to get the ethyl-6-alkyl-2,4-dihydroxybenzoate (IV). Add a solution of 24g NaOH in 200 ml water to (IV) and stir and reflux under nitrogen in hood for 3 hours. Cool in ice bath, acidify carefully with a cold solution of 20 ml concentrated sulfuric acid in 80 ml water while stirring under nitrogen in ice bath. Reflux under nitrogen 5 minutes, cool, extract 3 times with ether and dry, evaporate in vacuum the combined extracts to get about 80% yield of olivetol (or analog). The last step may not be necessary since (IV) may yield an active THC.

Dimethylheptylresorcinol CA 65,20062(1966)

This method is specifically designed to produce good yields of dimethylheptylresorcinol, which provides, after synthesis by any of the various routes, one of the most active THC analogs yet discovered. Note that the synthesis may not have to be carried all the way to the alkylresorcinol since the intermediate ketones etc. may give an active THC analog.

Mix 294g (1.6M) 1,3,5-trichlorobenzene, 184g (3.4M) Na methoxide and 450g (3.3M) diglyme and reflux at 162° C for 42 hours. Cool to room temperature, filter and distill the solvent to get 70% yield of 1-Cl-3,5-dimethoxybenzene (I). 43.2g (I) in 540 ml tetrahydrofuran is added dropwise to 7.3g Mg, a small crystal of iodine and a few drops of ethyl bromide (under nitrogen if possible) over ½ hour while the mixture is heated to 75° C. Reflux 2 hours and cool to room temperature to obtain the Grignard solution.

To anhydrous liquid ammonia in a cooled flask, add $Fe(NO_3)_3$. $9H_2O$ and then small pieces of sodium and bubble air through the solution until reaction is observed. Add more sodium portionwise until 11.5g is in solution. To the solution at its reflux temperature, add 27.6g (0.5M) proprionitrile (or isobutyronitrile etc. for analogs) and add the resulting solution dropwise to 53.3g 1-Cl-pentane (or 0.5M analog such as 2-Br-hexane etc.) and let the ammonia evaporate. Add ether, then water to the residue and separate the aqueous layer and extract with ether. Dry and evaporate in vacuum the combined ether solutions to give 64% yield of 2-methyl-heptanonitrile (II) (or analog).

Add 34.0g (0.27M) (II) in tetrahydrofuran to the Grignard solution of (I) over ½ hour, heat 6 hours at 60° C and hydrolyze with 1600 ml 50% sulfuric acid, keeping the temperature below 40° C. Evaporate in vacuum the solvent and add another 400 ml 50% sulfuric acid. Heat 1 hour at 95-100° C, cool, add ether, separate the aqueous layer and extract with ether. Dry and evaporate in vacuum the combined ether layers to get 71% yield 2-(3,5-dimethoxybenzoyl) heptane (III) (can distill 133-8/0.2), which can be demethylated as described for the preparation of (VII) below and possibly used to synthesize an active THC analog (as can IV, V, or VI).

21.8g (0.082M) (III) is added dropwise (keeping the temperature at 15-20° C) to 3 molar methyl-MgBr in ether and refluxed 1 hour. Pour into a sulfuric acid-ice mix, add more sulfuric acid and stir. Separate and extract the aqueous layer with ether. Wash, dry, filter and evaporate in vacuum the combined ether layers to get 2-(3,5-dimethoxyphenyl)-3-methyl-octanol (IV), which is dehydrated by mixing with anhydrous oxalic acid and heating to 130-40° C. Extract the dehydrated reaction products with benzene to get the octenes (V). Add 25g (V), 2.5g 65% Ni on kieselguhr in dry hexane to hydrogenator and hydrogenate at 1750 psi and gradually increase the temperature to 125° C. After 3 hours, increase the pressure to 1850 psi and hold there 2½ hours. filter and evaporate in vacuum to get 2-(3,5-dimethoxyphenyl)-3-methyl-octane (VI).

Add 20g (0.08M) (VI) to 38% HBr in glacial acetic acid and stir and reflux for 6 hours. Pour onto ice and water, neutralize with solid sodium carbonate and extract with ether. Extract the ether with 10% aqueous NaOH, acidify the aqueous solution with HCl, extract with ether and dry, evaporate in vacuum (can distill) to get 2-(3,5-dihydroxyphenyl)-3-methyloctane (VII) (5-(l,2-dimethylheptyl)-resorcinol).

As an alternative process for getting from (III) to (VI), combine 64.2g (0.18M) methyltriphenylphosphonium bromide in dry benzene with 11.6g (0.18M) (in 14% solution) butyllithium in benzene. Heat to 60° C and cool. 49.0g (0.176M) (III) in 40 ml dry benzene is added (keep temperature below 40° C) and then reflux 2 hours. Cool, filter and evaporate in vacuum to get the octene, which after catalytic hydrogenation as described for (V) yields (VI).

5-Alkylresorcinols Aust. J. Chem. 21,2979(1968)

Mix 50g 3,5-dihydroxybenzoic acid, 250 g K_2CO_3, 200 ml dimethylsulfate and one liter acetone and reflux 4 hours. Remove the acetone, add one liter water and one liter ether to the residue and extract. Wash the ether extract with 2X100 ml concentrated NH_4OH, 2X100 ml dilute HCl and 100 ml water and dry, evaporate in vacuum to get 48g methyl-3,5-dimethoxybenzoate (I). Recrystallize from aqueous methanol. To a stirred suspension of 19g lithium aluminium hydride in 200 ml ether add 78.4g (I) in 300 ml ether at a rate which gives gentle refluxing. Reflux 2½ hours, cool and add 50 ml wet ether; then 100 ml dilute sulfuric acid. Wash and dry, evaporate in vacuum the ether extract to get 62g oily 3,5-eimethoxybenzyl alcohol (II). Recrystallize from ether-pentane. To a cooled stirred slurry of 15g CrO_3 and 250 ml pyridine add 8.4g (II) in 25 ml pyridine and let stand 1 hour at room temperature. Add 60 ml methanol, let stand 2 hours, and dilute with 500 ml 5% NaOH and 500 ml ether. Extract the aqueous layer with ether and wash the combined ether layers with 500 ml water, 3X500 ml 5% sulfuric acid, 500 ml water and 200 ml saturated NaCl and dry, evaporate in vacuum to get 7g 3,5-dimethoxybenzaldehyde (III). Recrystallize from ether-pentane. To a flask with a dropping funnel and condenser add 0.58g Mg turnings and 10 ml ether. Add a few drops of a solution of lauryl bromide (5.7g) or equimolar amount of homolog in 15 ml ether and start reaction by adding 2 drops methyl iodide. Add the remaining bromide solution with stirring and gentle refluxing over 15 minutes and then reflux 3 hours. Cool in an ice bath and add 3.1g (III) in 5 ml ether dropwise with stirring over 45 minutes. Reflux 4 hours, cool and dilute with ice water. Wash the organic layer with 2X25 ml 3N sulfuric acid, 2X25 ml 10% K_2CO_3, 25 ml water, 25 ml saturated NaCl and dry, evaporate in vacuum to get 5g 3,5-dimethoxyphenyldodecyl methanol (IV) or homolog. Recrystallize from methanol. Hydrogenate 4.2g (IV) in 50 ml ethyl acetate with 5 drops concentrated sulfuric acid and 0.5g 10% Palladium-Carbon catalyst at room temperature and 5 atmospheres hydrogen for 4 hours. Filter and evaporate in vacuum to get the aklylresorcinol dimethyl ether.

Aust. J. Chem. 26,799(1973) gives a 2 step synthesis of 5-alkylresorcinols by condensation of beta-ketosulphones with 3,5-dimethoxybenzyl bromide and then reduction. Aust. J. Chem. 26,183(1973) gives a synthesis from 3,5-dimethoxy-N,N-dimethylbenzylamine in 7 steps (but perhaps only 4 will reach a cpd. that can give an active THC analog).

5-alkylresorcinols CPB 20,1574(1972)

To a solution of 0.02M ethyl- β -ketocaprylate (or homolog) in 20 ml tetrahydrofuran, add 1.02g (0.02M) (53% oil) NaH with stirring and cooling and then add a solution of diketene (1.68g, 0.02M) in 20 ml tetrahydrofuran dropwise, keeping the temperature between -5 and 0° C. Stir 1 hour at this temperature and then 1 hour at room temperature. Neutralize with 10% HCl and extract with ether. Dry and evaporate in vacuum to get about 38% yield of ethyl-olivetol carboxylate (I). (I) can be purified on silica gel, the impurities being eluted with petroleum ether (30-35° C) and the produce with 8:1 petroleum ether:ether. Recrystallize from n-hexane. Dissolve 0.2g (I) in 10 ml 10% NaOH and reflux 30 minutes. Acidify with 10% HCl and extract with ether. Wash the extract with water and dry, evaporate in vacuum (can distill 126-129/3) to get 96% olivetol (or homolog).

For new, simple, high yield syntheses of 5-alkyl resorcinols see TL 4839(1973), 2511(1975) and CJC 52:2136(1974). A superior, high yield, 3 step synthesis of olivetol and analogs has recently appeared in JOC 42:3456(1977).

MAGIC MUSHROOMS
AND OTHER INDOLE TRIPS

Various trees, vines, etc. and mushrooms containing dimethyl-tryptamine (DMT) and analogs have been used by the natives of Central and South America for millennia. Precise chemical and botanical identification have been made on a number of species, the first and most famous of these being certain species of mushrooms of the genus *Psilocybe*, species of which are found in the U.S.A., Canada, Scotland, Australia, etc. as well as Central America. Psychedelic species of the closely related genera *Conocybe*, *Stropharia*, *Pholiota*, *Copelandia* and *Panaeolus* are found widely scattered around the world. Not all species of each genus necessarily contain significant amounts of hallucinogens, but it seems that most do. A general test is that the stem of indole containing species tends to turn blue several hours after it is picked and slightly crushed. These are generally found in late summer and fall in the U.S.A. Pictures are given in many mushroom books including R. Heim, LES CHAMPIGNONS HALLUCINOGENES DU MEXIQUE (1958), and L. Enos, A KEY TO THE AMERICAN PSILO-CYBIN MUSHROOM (1971).

It turns out that the home cultivation of psilocybin mushrooms is quite easy. See Oss and Oeric PSILOCYBIN: MAGIC MUSH-ROOM GROWER'S GUIDE (1976 - And/Or Press) or THE COMPLEAT PSILOCYBIN MUSHROOM CULTIVATOR'S BIBLE (1976 - Hongero Press). Strains of *Stropharia cubensis* being grown on the West Coast are sufficiently strong that I have seen people who were very experienced with acid get higher than they had ever been on three fresh mushrooms. Remember that psilocybin is cross-tolerant with LSD, so you won't get off as well if you've done acid recently.

Puffballs of the genus *Lycoperdon* are also hallucinogenic, and activity has been claimed for *Boletus satana*, which occurs in the southeastern U.S.

In THE TEACHINGS OF DON JUAN, Don Juan seems to have taught Carlos to smoke the mushrooms, which might provide a different or heavier trip than ingesting them since they undoubtedly contain many compounds like DMT and 5-methoxy-DMT which are not orally active.

Puffballs seem to usually produce only auditory hallucinations. *L. marginatum* found over much of Europe and America is active (1 or 2 constitutes a dose).

New Guinea "mushroom madness" is apparently due to species of *Boletus, Russula* and *Heimiella*. See R. Heim NOUVELLES INVESTIGATIONS SUR LES CHAMPIGNONS HALLUCIN-OGENES (1967). Also see FIELD GUIDE TO THE PSILO-CYBIN MUSHROOM (available from P.O. Box 15667, New Orleans, LA 70175).

Psilocybin (4-phosphoryloxy-N,N-dimethyltryptamine) and psilo-cin (4-hydroxy-DMT) are among the active indoles in the mushrooms. Upon ingestion, psilocybin is hydrolyzed to psilocin; consequently there is no point in carrying the synthesis past the psilocin step. All the active naturally occurring compounds in this group seem to have the dimethylamine moiety, which is usually obtained in the course of chemical synthesis by using dimethylamine (DMA). If however, the DMA is replaced by diethylamine (DEA), dipropylamine (DPA), methylethylamine (MEA), pyrrolidine, etc., the potency and duration of action might be considerably increased. Likewise, an increased activity might be seen when the OH position of psilocin is replaced by methoxy or acetoxy. Human data are lacking for most of these compounds, but judging from animal experiments, the order of potency should be roughly as follows (MET is methylethyltryptamine, DET is diethyltryptamine):

4 or 5 acetoxy-MET > 4 or 5 MET > 4 or 5 methoxy-DET > 4 or 5
5-acetoxy-DMT

methoxy-DMT > 4-OH-DMT > DMT
DET

Compounds with a low relative activity (e.g., DMT, DET, 5 methoxy-DMT) have very little activity orally and must be smoked or sniffed. Unfortunately, these compounds taste and smell like burning plastic when smoked and are harder to smoke than hash. There is, however, no evidence for the notion that they are damaging. With the exception of DMT, DET, psilocin and psilocybin, most of these compounds are probably legal in most states.

Psilocin

In the case of the unsubstituted N, N-diakyltryptamines, duration of action increases as the chain gets longer in the order DMT, DET, DPT, and seems to decrease with further increase in length. The trip produced by a good dose of DMT typically lasts about one-half hour, whereas that for DPT can last three to four hours or more. There is some evidence that DET produces a better (i.e., more meditative and euphoric) trip than DMT.

If the alkyl side chain at the 3 position of the indole nucleus is shortened (e.g., gramine) or lengthened (e.g., 3-(3-dimethylamino)-propyl indole) activity seems to decrease strikingly. Also, as the substituents are moved around the benzene ring of indole, activity decreases greatly in the order 4,5,6,7. For example, whereas 4-OH-DMT(psilocin) is active at about 5 mg orally (i.e., about as active as STP), 5-OH-DMT(bufotenin) is not psychedelic at all.

The effects of 5-methoxy-DMT are unpleasant for most people (smoking it gives me nausea plus the feeling that I'm being sat on by an elephant), but it is not known whether other substituents in the 5 position or in the 4 and 5 positions simultaneously will be similarly distressing.

DMT, DET, etc. are remarkably fast acting (peaking in ca. 2 minutes after a good toke or snort) and produce very strong visual effects (my first toke of DMT produced a large grinning green dragon with red ruby eyes that lasted as long as the Stones' "Sympathy for the Devil"). It is unfortunate that it is usually DMT rather than the longer acting DET or DPT that is available, especially since the latter cpds. are no more difficult to produce. Also, it is rumoured that N,N-dibutyl and longer alkyls are not only active but (along with the dipropyl, diisopropyl etc. cpds.) orally active.

Ken Kesey has reputedly said that alpha-methyltryptamine, in oral doses of ca. 30 mg, peaks in about 12 hours, produces a trip similar to psilocybin, but nicer, and is the "Rolls Royce of psychedelics," but others find it unpleasant. Alpha-ethyltryptamine produces minimal LSD-type effects at 150 mg orally, but effects of these when smoked or inhaled are unknown. N,N-disubstituted tryptamines which have substituents in the alpha or beta positions should also be quite interesting.

Identification of Indoles
Keller Test

Add a little of the powdered substance (about 0.2 mg to 1 ml glacial acetic acid containing 0.5% $FeCl_3$, layer underneath with 1 ml concentrated sulfuric acid and shake. The color varies with the indole, being olive green for psilocin and red-violet for psilocybin.

Van Urk Test

Prepare Van Urk reagent by adding 0.5 g p-dimethylamino-benzaldehyde, 100 ml water, 100 ml concentrated sulfuric acid. Dissolve 1 mg substance in 1 ml ethanol and mix with 2 ml Van Urk reagent and illuminate for 10 minutes with an ultraviolet lamp (black light). Psilocin gives a blue-grey, psilocybin a red-brown color.

Colors produced in these two tests by many indole derivatives are given in HCA 42,2073(1959).

Quick, Easy, On-the-Spot Test

JPS 56,1526(1967)

Saturate strips of filter paper with 2% p-dimethylamino-benzaldehyde in 45% ethanol; air dry and store in tightly stoppered amber bottles (or keep in stoppered container in dark) which will keep them useful for several months. Put a little of the suspect substance in a few drops of ethanol (gin may do, but do a control), wet a filter paper strip in this and allow to dry. Put one drop concentrated HCl on the dried paper (don't let it touch anything). Alternatively, the powder can be placed directly on the strip and the HCl dropped on it. A violet red or violet blue spot indicates indole derivatives such as LSD. With DMT or psilocybin the color is redder. The color must be observed soon after adding the HCl since it rapidly changes.

Dialkyltryptamine Syntheses

Dialkyltryptamines HCA 42,2073(1959) and many others

To 25 g indole (or 50 g 4-benzyloxyindole or 0.21 M other indole) in 1 L dry ether at 0° add a solution of 50 ml (75 g) oxalyl chloride in 1 L dry ether carefully and with good stirring a little at a time over ½ hour and stir until bubbling ceases (about one-half hour more). Some indoles require a longer reaction time (e.g., 4-Cl-indole requires fifteen hours refluxing) and some will not react (e.g., 4-Br-indole). Add portionwise, carefully with stirring at 0°, a solution of

225 ml (160 g) diethylamine (DEA) (or 0.46M dipropylamine, pyrrolidine, etc.) in 100 ml dry ether at 0°. Stir and let warm to room temperature; cool, filter, and wash precipitate two times with ether to get (I). This can be recrystallized by dissolving in the minimum volume of 1:1 methanol:benzene (or 95% ethanol), gently heated, cooled to 0° and filtered (or add petroleum ether to induce precipitation). Dissolve 25 g (or 0.102 M) (I) in the least volume (about 200 ml) THF and add very carefully and slowly (preferably dropwise) to 20 g lithium aluminum hydride dissolved in the least volume (about 200 ml) tetrahydrofuran at room temperature. Stir and heat under reflux for about fifteen hours. Cool to 0° and slowly and carefully add a little cold methanol and water until no more bubbles are formed. Filter, wash precipitate with hot tetrahydrofuran and add washings to filtrate. Dry, evaporate in vacuum the tetrahydrofuran (or add petroleum ether) to precipitate the dialkyltryptamine. To purify, add 500 ml saturated sodium sulfate and filter. Wash precipitate with tetrahydrofuran; acidify with a few ml 0.1 M HCl and shake with ether. Separate the organic layer and neutralize with 0.1 M NaOH. Extract with $CHCl_3$ and dry, evaporate in vacuum the extract (or can evaporate until a few ml left and precipitate by adding petroleum ether). The 4-benzyloxy-DET which would be produced if 4-benzyloxyindole is used as the starting material is probably a good psychedelic. If however, it is desired to change this to 4-OH-DET, add 37.5 g 4-benzyloxy-DET in 1.2 L methanol to 20 g 5% Palladium catalyst on alumina (or 14 g 10% Palladium-Carbon) with 2.8 kg/cm^2 H_2 in a Parr hydrogenator and shake twelve hours. Filter, evaporate in vacuum. Other hydrogen-ating methods might also split off the benzene ring. Other methods (LAC 576,69(1952)) must be used for reducing a methoxy group to a OH group (another demethylation method is given here later).

If (I) has an alkyl group in position 1 (as in some of the following syntheses), reduction will give the indolylhydroxylamine. This may be active, but if the indolamine is desired (I) (substituted or not) may be reduced with the diborane method given later in this section.

Dialkyltryptamines JCS (C) 2220(1967)

This procedure gives about 20% yield with indole, but the yield with substituted indoles (e.g., 4-OH-indole for producing psilocin) has not been reported.

Cool 32 g ethyl iodide to 0°; dissolve in 50 ml anisole (other solvents won't work) and add 8 ml to 5.28 g Mg turnings in 50 ml anisole, and add the rest gradually. Warm gently to start the reaction, and if necessary add a crystal of iodine or a small amount of ether for a rapid rate. Stir well and heat at 50-60° for one hour (under N_2 if possible). Cool to 10° and add dropwise over one-half hour 12 g (0.1 M) indole in 50 ml anisole (keep temperature below 25°). Stir forty-five minutes at 50° and cool to -5°. Finely grind 0.2 M (34 g) 1-Cl-2-diethylaminoethane-HCl (or the corresponding diisopropyl, pyrrolidyl, etc. compounds) and suspend in about 20 ml benzene at 0°. The free base in benzene can also be used, if obtainable. Stir and take pH to 8.5 with 40% NaOH. Add anhydrous potassium carbonate until the water layer is semisolid. Decant the benzene and extract the residue with 4X15 ml benzene. Dry the combined benzene extracts with KOH pellets for less than an hour and quickly proceed to the next step. Add the benzene solution (about 80 ml) slowly over one hour to the above solution of indole in anisole at -5°. Stir three hours at -5° and let sit five hours at -5°. Then let warm to room temperature and dry, evaporate in vacuum (or to purify, break up the precipitate and pour the solution on 500 ml saturated aqueous NH_4Cl. Stir one-half hour; separate the organic layer and extract the aqueous layer with ether. Combine the organic solutions and extract three times with 10% HCl. Wash HCl extract with ether; cool to 0°, basify with 40% NaOH and extract three times with ether. Dry, evaporate in vacuum this second ether extract to get the oily DET or analog).

4-Substituted Dialkyltryptamines CT 279(1970)
(cf. JOC 30,339(1965))

Beta-carbomethoxypropionyl chloride (Org. Synth. 25,19 (1945)). Dissolve 400 g succinic anhydride in 190 ml methanol in 1 L round bottom flask and reflux (steam bath) one-half hour. Stir until homogeneous (about twenty minutes) and reflux one-half hour. Evaporate in vacuum and cool the residual liquid to precipitate about 500 g methyl-hydrogen succinate (I). Dissolve 264 g (I) in 200 ml $SOCl_2$ in a 1 L round bottom flask with a reflux condenser and warm at 30-40° in water bath for three hours. Evaporate in vacuum the $SOCl_2$ (can heat flask in steam bath) to get 270 g of the title compound (can distill 92/18).

Add excess diethylamine to beta-carbomethoxypropionyl-Cl in dry ether to get (see above JOC reference) 3-carbo-methoxy-N,N-diethylpropionamide (I). 206 g (I) in 3 L, three-necked round bottom flask in ice bath with stirrer, dropping funnel and reflux head. Keep temperature at 10-20° and add 169 g $POCl_3$ dropwise over fifteen minutes. Remove ice bath and stir fifteen minutes and replace ice bath. Add 250 ml ethylene chloride, cool to 5°. Stir and slowly add 67 g pyrrole in 250 ml ethylene chloride over 1 hour. Remove ice bath and reflux fifteen minutes (HCl evolution). Cool to room temperature and add solution of 750 g sodium acetate trihydrate in 1 L water dropwise at first, then as rapidly as possible. Reflux fifteen minutes with stirring, cool and remove ethylene chloride in separatory funnel. Extract aqueous phase with 3X200 ml ether and wash combined ethylene chloride and ether with 3X100 ml saturated aqueous Na carbonate (add carefully at first). Dry, evaporate in vacuum the organic phase to get 132 g methyl (pyrrolyl-2')-4-keto-4 butyrate (II) (can distill 135-45/0.2) (recrystallize from cyclohexane). Alternatively, (Chem. Commun. 1429(1968)), condense 1,3-cyclohexanedione and aminoacetaldehyde dimethylacetal in benzene with p-toluenesulfonic acid. Azeotropic removal of water gives a compound which, when treated with 3N HCl, gives compound (IV). But it has been claimed that this alternative method does not work.

94 g (II), 1.5 L diethylene glycol, 93 g hydrazine hydrate and heat at 100° fifteen minutes. Add 150 g potassium carbonate a little at a time and raise temperature slowly. Heat four hours at 190-200° and pour onto 5 kg ice. Acidify and then extract with 5X400 ml ether. Wash extract with a little saturated NaCl and evaporate in vacuum to get about 48 g 4-(pyrroly-2')-butyric acid (III) (recrystallize-cyclohexane). Can purify the oily compound by filtering through 80 g silica and 80 g Celite (elute with benzene). 24 g (III) in 400 ml 1,2-dichloroethane; cool to -5° and add 15.8 g triethylamine and 17 g ethyl chloroformate. Let stand 1 hour at 15°; filter and wash precipitate with 100 ml dichloroethane. Add 0.5 L anhydrous $ZnCl_2$ (freshly fused) and let stand two hours at -5°. Add 0.5 L 2N HCl, decant and wash the aqueous phase three times with $CHCl_3$; dry, evaporate in vacuum (can chromatograph as for (III)) to get about 11 g 4,5,6,7-tetrahydro-4-indolone (IV). 7.35 g (IV), 5 g 10% palladium-carbon, 700 ml mesitylene and reflux eight hours or more. Filter hot, wash precipitate with methanol, cool and evaporate in vacuum (can chromatograph as for (III)) to get 5 g 4-OH-indole (V) (recrystallize from petroleum ether), which can be converted to the diakyltryptamine by any of the methods described here or as follows (the first step leading to (VI) may not be necessary). 5 g (V), 20 ml pyridine, 10 ml acetic anhydride and heat in water bath 10 minutes. Pour on ice, stir and add $NaHCO_3$. After one-half hour extract with ethyl acetate, wash

extract with NaCl and dry, evaporate in vacuum to get about 6.3 g 4-acetoxy-indole (VI). 6 g (VI), 150 ml ether; cool in ice-salt bath and add carefully 6 ml oxalyl-Cl. After four hours add 20 g dry dimethylamine or equimolar amount other amine and stir twenty hours. Filter, wash precipitate with ether and then water to get about 2.8 g 4-acetoxy-3-indolyl-N, N-dimethylglyoxylamide (VIIa) (recrystallize-isopropanol). Shake the ether with water and filter to get about 5 g of the 4-OH compound (VIIb) (recrystallize-isopropanol). 7.8 g (VIIa or b or mixture obtained by evaporating in vacuum the ether above), 17 g lithium aluminum hydride, 150 ml tetrahydrofuran or dioxane; reflux seventeen hours, carefully add water and stir until bubbling ceases and evaporate in vacuum to get about 4.7 g of psilocin or analog (about 5% overall yield). For other methods of synthesizing (IV) see JOC 36,1232(1971) and references therein. For another method of reducing (IV) see Chem. Het. Cpds. (Russian), 572(1972).

4-Substituted Dialkyltryptamines HCA 42,2073(1959), 38,1452 (1955), CT 276(1970)

Method is illustrated for 4-benzyloxyindole (I) but will probably work for most other substituted indoles.

A: Convert (I) to 4-benzyloxygramine (II) as described elsewhere here.

B: Add 30 g (II) over one-half hour to 420 ml methyl iodide and let stand fifteen hours at 5°. Separate the iodomethylate which precipitates, dry briefly at 50° and heat with vigorous stirring at 80° for two hours with 60 g NaCN in 1 L water. Extract with CHCl₃, dry and evaporate in vacuum the extract and dissolve the residue in 250 ml ether. Filter, evaporate in vacuum to a few ml and precipitate the acetonitrile (III) by adding petroleum ether. The acetonitrile can also be prepared directly from the indole via the Grignard reagent as given elsewhere here.

B (Alternative): 0.05 M (II), 0.76 ml glacial acetic acid in 75 ml tetrahydrofuran (dry) are added slowly with stirring and cooling over one-half hour to a solution of 25.2 ml dimethylsulfate and 0.76 ml glacial acetic acid in 30 ml dry tetrahydrofuran. After two hours, filter and wash the precipitate with ether. Dissolve precipitate in 10% aqueous solution of KCN and heat one hour at 70°. Filter, wash precipitate with water and dry to get (III).

C: 0.04 M (III) in 200 ml 33% ethanol solution of DMA or other amine, 2.5 g Raney-NI, 40°, 100 kg/cm² (about 100 atmospheres)

H$_2$. Heat about three hours; filter and evaporate in vacuum to get the dialkyltryptamine.

C (Alternative): 5.8 g (III), 12 g KOH, 36 ml ethanol, 28 ml water; reflux fifteen hours, add 15 ml glacial acetic acid, filter and add 150 ml water to precipitate 4-benzyloxyindole acetic acid (IV). Filter, wash precipitate with water and recrystallize from methanol.

D. 1.76 g (IV), 1.4 g PCl$_5$, 50 ml ether at 0°. Stir until dissolved and add dropwise to solution of 5.36 g DEA (or equimolar amount other amine) in 10 ml ether. Let warm to room temperature, let stand one-half hour and precipitate by adding water. Filter, dry, evaporate in vacuum the ether and add the residue to the precipitate to get the diethylacetamide (V) (recrystallize-benzene).

D (Alternative): 20.6 g (IV) in 50 ml methanol; add excess diazomethane in ether, evaporate in vacuum and dissolve the oil in 90 ml dry hydrazine. Heat at 135° 1½ hours, add 150 ml water and cool to precipitate the hydrazide (recrystallize-aqueous methanol). 14.7 g of the hydrazide in 250 ml tetrahydrofuran or dioxane and add 50 ml 1N NaNO$_2$. Cool to 4° and add dropwise over 4 minutes with vigorous stirring, 60 ml 1N HCl; let stand fifteen minutes at 4° and add 500 ml water. Extract the oily azide with ether and dry, evaporate in vacuum. Add 77 ml (0.75 M) DEA (dry) to the azide and let stand three hours at 5° with care to exclude moisture. Evaporate in vacuum and take up the residue in NaHCO$_3$. Extract with CHCl$_3$ and dry, evaporate in vacuum the extract to get (V).

E. 0.7 g (2.28 mM) (V) in 20 ml dry tetrahydrofuran; add slowly to a well stirred solution of 0.35 g lithium aluminum hydride in 20 ml tetrahydrofuran and keep one hour at 40°. Carefully add 5 ml water and stir twenty minutes. Add 15 ml 20% NaOH and extract with ether. Dry, evaporate in vacuum the extract to get 4-benzyloxy-DET or analog (recrystallize-ether).

E (Alternative): To 0.4 g (V) in 15 ml tetrahydrofuran add 2.9 ml 1 M borane in tetrahydrofuran and reflux one hour. Cool and heat with 5 ml 2N HCl; evaporate in vacuum to get about 0.15 g product.

4-Nitro and 4-Amino-Dialkyltryptamines CJC 41,2585(1963)

153 g alpha-Cl-butyryl-Cl, 16 g Pd on BaSO$_4$, 1.66 ml sulfur-quinoline (Org. Synthesis 21,84(1941)), 900 ml toluene; reflux and stir while bubbling H$_2$ through for seven hours or until HCl evolution ceases (can bubble effluent HCl through water to monitor evolution).

Filter, wash toluene with water, $NaHCO_3$, water and dry, evaporate in vacuum to get 100 g gamma-Cl-butyraldehyde (I) (can distill 28/2). 10 g (I), 20 g 3-nitro-phenylhydrazine; dissolve in the minimum volume of hot ethanol containing 10% glacial acetic acid. Heat on steam bath one hour; cool and add water until dark oil separates. Evaporate in vacuum the ethanol and decant the water to get the oily gamma-Cl-butyraldehyde-3-intro-phenylhydrazone(II). 29 g (II), 300 ml concentrated HCl, 200 ml benzene; stir three hours, replace benzene with fresh benzene and stir four hours. Combine the two benzene portions, wash with water and dry, evaporate in vacuum to get 4 g 3-(beta-Cl-ethyl)-4 and 6-nitroindole (III). 3.56 g (III), 200 ml ethanol, 200 ml 34% aqueous DMA (or other amine) and let stand at room temperature for one week. Evaporate in vacuum the ethanol, filter, dissolve the precipitate in dilute HCl and filter. Basify the filtrate with dilute NaOH to precipitate 3 g 4 and 6-nitro-DMT (IV). 5.2 g (IV), 350 ml ethanol, 100 ml 1N NaOH; heat to 50° and add a solution of 3 g Na dithionite in 15 ml 0.2N NaOH. filter hot and evaporate in vacuum to get 2 g 4 and 6 amino-DMT (can purify by dissolving in HCl, filter, basify, extract with ether and dry and evaporate in vacuum the extract).

5-Acetyl-DMT JMC 7,144(1964)

40 g p-aminoacetophenone, 250 ml water, 143 ml concentrated HCl. Slowly add 21 g $NaNO_2$ in 200 ml water and keep temperature at 0-5°. Add 70 g ethyl-alpha-(2-dimethyl-amino-propyl)-acetoacetate, then 63 g sodium acetate and keep pH at 5.5-6.0 with 3N NaOH. Stir in cold two hours; basify with NaOH and extract with 3X400 ml $CHCl_3$. Dry, evaporate in vacuum the extract to get 70 g ethyl-alpha-keto-delta-dimethylaminovalerate (I) (recrystallize-benzene-petroleum ether). 50 g (I), 430 g polyphosphoric acid and heat slowly with stirring. Foaming starts about 60°. Slowly raise temperature to 105° and keep two hours. Cool to 70°; pour into 700 ml ice water; stir to dissolve; cool, basify, extract with 3X400 ml $CHCl_3$ and dry, evaporate in vacuum to get 10 g 5-acetyl-2-carbethoxy-DMT (II). Test for activity. 11.2 g (II) in 190 ml 20% HCl; reflux four hours; cool and filter. Basify with 40% KOH and extract with 4X125 ml $CHCl_3$ and dry, evaporate in vacuum to get 5-acetyl-DMT.

1-Methyl-DET BSC 1056(1962)

135 g acetoacetic acid in two-necked flask fitted with two

condensers atop one another, the upper air cooled only, and a dropping funnel with a tube running to the bottom of the flask. Heat to boiling and add over forty minutes, as a vapor, 80 g DEA dried (e.g., with Na wire). The residue can be distilled (120-140/12) and redistilled (123-4/12) to give 100 g N,N-diethylacetoacetamide (I). 15.7 g (I) in 80 ml $CHCl_3$ and add dropwise 16 g Br_2 in 10 ml $CHCl_3$. Heat gently to a boil for one-half hour and cool to precipitate. Filter, wash precipitate with $CHCl_3$ and dry to get 25 g gamma-Br-diethylacetoacetamide (II) (use crude since decomposes on distillation). 4.72 g (II), 4.28 g N-methyl-aniline, 20 ml dimethyl formamide and let stand twelve hours (or 90 ml ethanol and reflux eighteen hours) at room temperature. Slowly add 300 ml water and extract the oil which forms with benzene. Wash with water and dry, evaporate in vacuum the benzene extract to get 4 g precipitate (recrystallize-80% ethanol). 4 g precipitate, 4 g $ZnCl_2$ finely ground; heat in oil bath and keep temperature 100-110° for forty-five minutes. Cool and dissolve precipitate in 40 ml 4N HCl and 160 ml benzene. Separate the benzene and wash with water; basify and dry, evaporate in vacuum to get 1.3 g 1-methyl-3-indole-N, N-diethylacetamide (III) (recrystallize-ethanol). Test for activity. Recover N-methyl-aniline by basifying the water or ethanol, extract precipitate with ether, wash extract to neutrality and dry, evaporate in vacuum. 1.1 g (III), 0.38 g finely ground lithium aluminum hydride, 300 ml ether and reflux two days. Carefully add a little water and filter, evaporate in vacuum to get 1-methyl-DET (recrystallize-ethanol). (III) can probably also be reduced by the method described in the chemical hints section or even more simply as follows: Dissolve 1M $NaBH_4$ and 0.1 M (III) in 500 ml pyridine or other solvent and reflux eight hours or more.

Alpha-alkyl-DMT TET 29,971(1973)

21.6g(0.1M) alpha-bromopropionyl (or butyryl etc.) bromide is added dropwise over 1 hour to a well stirred mixture of 11.7 g (0.1M) indole and 8.1 ml (0.1M) pyridine in 300 ml toluene at 60°. Stir 1 hour, cool and pour into 500 ml water. Separate the oil and dissolve in methanol. Let stand 1-24 hours until crystals separate. Filter (recrystallize from acetonitrile) to get 18.4 g (72%) 3-(2-bromopropionyl)indole(I). 5.2 g (0.02M) (I), 7 ml 33% aq. dimethylamine and 3 g NaI in 100 ml ethanol are refluxed for 20 hours, concentrated to 25 ml and poured into 200 ml aqueous 0.5M HCl. Extract with ether and basify with concentrated NH_4OH. Recrystallize from ethanol to get ca. 3g(50%) 3-(2-dimethylaminopropionyl)indole(II). 2.7g (II) in 50 ml tetrahydrofuran is added to a well stirred mixture of 2.7 g lithium aluminum hydride in 60 ml tetrahydrofuran. Reflux 23 hours, carefully add 5 ml 2N KOH, filter and wash the the ppt. with ether and dry, evaporate in vacuum the ether to get 1.65 g (66%) alpha-methyl-DMT (recry.-benzene-n-hexame).

Alpha-methyl-DET JMC 9,343(1966)

46.8 g (0.4 M) indole in 100 ml toluene; add to 54.5 g ethyl-bromide and 12.5 g Mg turnings in 125 ml ether. After one-half hour convert the indoyl-Mg-Br to 3-indoyl-2-propanol with pro-pylene oxide (CA 56,3455(1962)). 8.8 g of the indolyl-propanol, 200 ml ether; add 4.4 g PBr_3 and let stand four hours. Add excess DEA and stir for a few minutes. Evaporate in vacuum or extract with dilute HCl and basify the extract with NaOH to precipitate the alpha-methyl-DET.

Dialkyltryptamines from Tryptamines BCSJ 11,221(1936)

Illustrated for 5-methoxy-tryptamine (I). 1.5 g (I), 30 ml ethanol; add 5 g methyl iodide (or equimolar amount ethyl iodide) and 4.5 g dry sodium carbonate and heat five hours on water bath. Filter hot, heat precipitate with ethanol and filter hot again. Evaporate in vacuum to get 2.5 g 1-methyl-5-methoxy-DMT.

Dialkyltryptamines AP 294,486(1961)

Convert indole to indolyl-3-methyl-ketone (I) by treating indolyl-Mg-Br (preparation already described) with acetyl-Cl, by treating indole in $POCl_3$ with dimethylacetamide (Vilsmeier reaction), or by reacting indole with diketene (ACS 22,1064(1968)). 15.9 g (I) in 50 ml methanol; cool, stir and add dropwise 16 g Br_2. Reflux 1½ hours on water bath; cool, filter, wash with ether and recrystallize-methanol to get 18 g indolyl-3-Br-methyl-ketone (II). Dissolve 11.9 g (II) in 60 ml warm isopropanol and add 11 g 38% aqueous DMA (or equimolar amount other amine); reflux one hour on water bath. Filter (recrystallize-ethanol) to get 8.5 g indolyl-3-dimethylamino-methyl ketone (III). Add 4.6 g (0.02 M) (III) in 30 ml tetrahydrofuran to 2.3 g lithium aluminum hydride in 50 ml tetrahydrofuran, stir one-half hour at room temperature and reflux two hours. Add a little water dropwise and extract the precipitate with acetone. Dry, evaporate in vacuum the combined organic phases to get an oil which will precipitate with ether-petroleum ether to give DMT. (III) should be tested for psychedelic activity.

Dialkyltryptamines BCSJ 11,221 (1936), BSC 2291 (1966)

30 g 5-methoxy-indolyl-3-acetonitrile is heated with KOH in aqueous methanol until no more ammonia is evolved (about 20 hours). Evaporate the methanol in vacuum and extract the water remaining with ether. Acidify the aqueous layer with HCl to precipitate 28 g 5-methoxy-tryptophol (I). Alternatively, dissolve 2.3 g indole in 15 ml glacial acetic acid and 5 ml acetic anhydride. Add with stirring 0.025M ethylene oxide, heat to 70° for 25 hours, then hold at 20° for 50 hours in a closed flask. Pour into water and extract with ether. Wash with water, dry, evaporate in vacuum and saponify (e.g., heat with NaOH) the residue to get 1.5 g tryptophol (can purify on alumina; benzene elutes indole, ether elutes tryptophol). 2g (I) or tryptophol in 100 ml ether. Mix with 1 g PBr_3 dissolved in ether and let stand 12 hours at room temperature. Decant the liquid from the precipitate; wash with water and $NaHCO_3$ and dry, evaporate in vacuum the ether to get 1.3 g of the oily bromide (II). 1 g (II), 4 ml methanol, 4 ml 33% aqueous DMA (or DEA etc.) and heat on steam bath in sealed container 15 hours. Acidify with about 50 ml dilute HCl, extract with ether and dry, evaporate in vacuum the ether to get about 0.5 g DMT or analog.

Dialkyltryptamines BSC 1335(1966)

To 600 ml liquid NH_3 add 23.5 g Na: 40 g Na-amide are thus prepared and the NH_3 evaporated in vacuum. Mix 170 g 5-Cl-2-methoxy-phenylacetonitrile (preparation given elsewhere) in 900 ml benzene with the Na-amide and stir and reflux two hours. Cool to 40° and add dropwise 111 g 2-dimethylaminochloroethane (or diethyl etc. analogs) prepared freshly in benzene as described in a previous method, or use the base freshly distilled. Reflux two hours, cool and add a little ethanol and water and extract the amine by evaporating in vacuum or pouring on cool water and filtering to get 157 g (I) (recrystallize-petroleum ether). 105 g (I) in 150 ml methanol containing 15% NH_3. Hydrogenate at 50°, 70 kg pressure, in presence of Raney-Ni. Filter, dry and evaporate in vacuum to get 97 g of the phenethylamine (II). Cyclise (II) with Na and naphthalene as described later for 4-methoxyindole to get the yellow, oily indoline (III) (recrystallize-ethanol). Test this for activity. If desired, the noncyclised material can be eliminated by tosylation. 3.1 g (III), preferably as the HCl salt, 200 ml water; stir and heat fifteen hours with 10 g Raney-Ni and 1.5 g maleic anhydride. Filter, dry and evaporate in vacuum to get 4-methoxy-DMT.

Dialkyltryptamines

A: CCCC 22,1848 (1957). 5.5 g indole, 15 ml cyclohexan, ½ g

copper. Reflux and add dropwise 2.9 g diazoacetone. After a time, the reaction goes very rapidly and forms two layers. Filter, evaporate in vacuum or distill (130-145/0.2) to get 2.6 g 3-indolyl-acetone (I).

A (Alternative): JCS 3175(1952). 2 g 3-indolyl-acetic acid (preparation given elsewhere here), 1.55 g freshly fused sodium acetate, 5 ml acetic anhydride. Heat 135-140° on oil bath for eighteen hours; cool, wash with water and extract with $CHCl_3$-ether (1:4). Wash organic phase with 3X20 ml saturated $KHCO_3$ and dry, evaporate in vacuum to get the 1-acetyl-3-indolyl-acetone, which can be reduced to the alpha-methyl-tryptophol derivative with lithium aluminum hydride, and then converted to the dialkyl-tryptamine as already described (as can (I)), or used in step B, or reduced to (I) as follows: dissolve 1 g in 1 ml 1 N Na-methoxide in methanol and 60 ml methanol, and keep at 40° for 10 minutes: acidify with dilute HCl and extract with ether. Dry, evaporate in vacuum to get (I) (recrystallize-methanol).

A (Alternative): JCS 2834(1962). 4.8 g Mg turnings, 32 g ethyl iodide in 20 ml dry anisole. Cool to 0° and add dropwise 15.6 g indole in 20 ml anisole. Stir one-half hour at 20°; cool to 0° and treat with 20 ml prop-2-ynyl-bromide in 10 ml anisole over 20 minutes. Continue stirring one hour at 0° and let stand at room temperature twelve hours. Cool to 0°, add 100 ml ether, 200 ml water, 12 ml glacial acetic acid, 100 ml water, and extract with 5X25 ml ether. Wash with $NaHCO_3$ and dry, evaporate in vacuum the extract to get 8 g oily material which precipitates on standing in refrigerator. Add ½g $HgSO_4$ to 100 ml 2N sulfuric acid; stir and heat on steam bath and add 15 g of the precipitate in 100 ml ethanol. Stir and heat two hours and pour into water. Basify with $NaHCO_3$ to get 3 g (I) (recrystallize-benzene).

B: JMC 9,343(1966). 3.3 g (I) in 100 ml ethanol; reduce over palladium-carbon catalyst in presence of 0.04 M DEA (or other amine). After two hours filter and evaporate in vacuum to get the DET or analog.

Methylethyltryptamines HCA 49,1199(1966)

82 ml acetic anhydride, 35 ml formic acid; heat two hours at 50-60° with stirring and then add dropwise a solution of 100 g 4-benzyloxytryptamine (or equimolar amount other tryptamine) in 250 ml tetrahydrofuran. Cool to room temperature and stand twelve hours. Evaporate in vacuum, add 100 ml water and let stand two hours to precipitate the oily N-formyltryptamine (I) (recrystallize-700 ml 50% ethanol). 35 g (I) in 150 ml tetrahydrofuran; add

51

dropwise to 20 g lithium aluminum hydride in 400 ml tetra-hydrofuran and reflux twelve hours. Cool to 15°, slowly add 35 ml ethyl acetate and reflux two hours. Slowly add 10 ml water, 10 ml 15% NaOH, 30 ml water. Filter, wash precipitate with tetra-hydrofuran and evaporate in vacuum to get 20 g 4-benzyloxy-methylethyltryptamine (II) or analog (can recrystallize by dissolving in 100 ml methanol and precipitating with 10 g oxalic acid (anhydrous)).

Alternatively, dissolve 220 g 4-benzyloxy-3-indoleacetic acid (or equimolar amount other indoleacetic acid) in 2 L absolute methanol and reflux six hours in the presence of 20 g Dowex 50X8 sulfonic acid resin. Filter (decolor with carbon if desired) and concentrate below 35° until precipitation starts; then cool to precipitate and filter to get 200 g of the methyl ester. Add 200 g of the ester to 600 ml 40% aqueous methylamine over twelve hours with vigorous stirring. Filter, wash precipitate with water and dry to get 187 g of the N-methyl-acetamide (reflux two hours in 500 ml benzene to remove unreacted ester). 24 g of the acetamide in 300 ml tetrahydrofuran is added dropwise to 10 g lithium aluminum hydride in 300 ml tetrahydrofuran; reflux ten hours, cool to 15° and add dropwise with stirring 50 ml ethyl acetate. Reflux two hours and proceed as above to get 15 g (II) or analog.

Dialkyltryptamines JACS 84,4917(1962)

Illustrated for the piperidine analog of DMT.

2.5 g methyl or ethyl-indoleacetic acid (see previous method and elsewhere for preparation), 12 ml dry piperidine and reflux twenty-four hours. Evaporate in vacuum, dissolve residue in $CHCl_3$; wash with dilute HCl, $NaHCO_3$, and dry and evaporate in vacuum to get 2.6 g 3-indolyl-aceto-piperidide (I) (test for activity). 2.6 g (I) in 80 ml tetrahydrofuran; add 2.5 g lithium aluminum hydride in 200 ml ether and reflux five hours. Carefully add a little water until bubbling stops. Filter, dry and evaporate in vacuum to get the piperidine analog of DMT.

Tryptamine from Tryptophan JCS 3993(1965)

10 g tryptophan in 500 ml diphenyl ether; reflux one hour under N_2; cool and extract with 3X40 ml 2N HCl. Wash extract with ether, basify with 6N NaOH and extract with 5X50 ml ether. Wash extract with water, NaCl; dry and evaporate in vacuum and recrystallize-benzene to get tryptamine which can be alkylated to dialkyltryptamine as described previously. (Tryptophan, if oxidized

by H_2O_2 and cyclised with HCl, gives a psychedelic compound.) Much interst has been generated by the mention of tryptophan oxidation to a psychedelic compound, but the original reference gives no details of the method.

Dialkyltryptamines CCCC 24,3984(1959)

A: 4 g 3-indoleacetic acid (I) in ether. Treat with PCl_5 as described elsewhere here to get 2.7 g of the chloride (II). Dissolve 2.7 g (II) in 40 ml ethyl acetate and add 2 ml piperidine (or equimolar amount other amine), 3.5 ml N-ethyl-piperidine in 40 ml ethyl acetate. Let stand three hours at room temperature and filter. Wash filtrate with 1N HCl, 10% Na carbonate and evaporate in vacuum to get 1 g of the amide (III) (test for activity). Add 1 g (III) in 25 ml ether to 1.4 g lithium aluminum hydride in 50 ml ether; stir four hours at room temperature and reflux one-half hour. Cool and carefully add water until no more bubbling. Add 3 ml 20% NaOH and dry, evaporate in vacuum to get 1 g piperidine analog of DMT (IV).

A (Alternative): Treat 4 g (I) in pyridine with $SOCl_2$ as described elsewhere here to get (II). Treat (II) with 2.3 g triethylamine and 2 g piperedine (or other amine) in 5 ml ether. Stir fifteen hours at room temperature; extract with 200 ml water, and after precipitation, wash precipitate with 10% Na carbonate, 3N HCl and dry, evaporate in vacuum the ether solution to get (III). Reduce (III) to (IV) as above or with the $NaBH_4$ method or possibly with hydrogenation.

A (Alternative): To 6 g (I) in 250 ml ether add 3.1 g piperidine (or other amine) in 20 ml ether. Let precipitate form and filter. Heat 7 g precipitate three-and-one-half hours at 190-215°; cool and dissolve in 100 ml ether. Wash with 10% K carbonate and 5N HCl and evaporate in vacuum to get (III), which is reduced to (IV) as above.

Dialkyltryptamines JCS 7175,7179(1965)

6.3 g N-Methyltryptamine in 40 ml methylformate; heat in autoclave six hours at 100° and evaporate in vacuum to get 7.4 g precipitate (test for activity). Dissolve precipitate in 50 ml tetrahydrofuran and add to 2.6 g lithium aluminum hydride in 50 ml tetrahydrofuran and reflux three hours. Carefully add a little water or methanol; filter, evaporate in vacuum to get DMT or analog. This paper also gives four ways to prepare hydroxylamine analogs of DMT, but their activity is unknown.

If 3-methoxy-aniline is used, 6-methoxy-DET will probably result, but if 3,5-dimethoxyaniline is used, 4,6-dimethoxy-DET should be obtained.

A: 1M p-methoxy-aniline (p-anisidine), 0.5 M ethyl-gamma-Br-acetoacetate (brominate as described above for 1-methyl-DET synthesis); cool and add 250 ml ether. Filter, evaporate in vacuum and reflux residue fifteen hours with 60 g $ZnCl_2$ in 250 ml ethanol. Evaporate in vacuum, wash precipitate with water and dissolve residue in benzene. Wash with 4N HCl and water and dry, evaporate in vacuum. Reflux prcipitate two hours in ethanol-KOH to get about 70% yield of 5-methoxy-indoleacetic acid (I).

B: Dissolve 4.26 g triethylamine and 8 g (I) in 200 ml $CHCl_3$; cool to -5°, add rapidly 4.58 g ethyl-Cl-carbonate and agitate fifteen minutes. Bubble dry NH_3 through for five minutes (can possibly substitute concentrated NH_4OH) and let stand one hour at room temperature. Add water, wash with $NaHCO_3$ and water and dry, evaporate in vacuum the $CHCl_3$ to get 5-methoxy-diethyl-indolyl-acetoacetamide(II). Test this for activity.

B (Alternative): 0.5 g (I) and 0.28 g triethylamine in 5 ml methylene Cl. Add 0.33 g dicyclohexylcarbodiimide and let stand for sixteen hours at room temperature. Add a little glacial acetic acid, wash with dilute HCl, NaHCO and water and dry, evaporate in vacuum to get (II).

B (Alternative): Do not reflux the precipitate in ethanol-KOH in the last part of step A. Heat 8 g of the precipitate with equimolar amount DEA (or other amine) in 80 ml methanol in a sealed bomb or tube in autoclave for twenty-four hours at 105° and filter, evaporate in vacuum to get (II) (recrystallize-ethanol).

B (Alternative): 20 g of the unrefluxed precipitate (ethyl-ester of (I)) from last part of step A in 100 ml ether. Add dropwise to a solution of 4 g lithium aluminum hydride in 900 ml ether at 0°. Reflux three hours and isolate the resulting tryptophol as described earlier. Dissolve 3 g of the tryptophol in 140 ml ether and stir at 0°. Add dropwise 1.8 g PBr_3 in 30 ml ether and let stand sixteen hours at room temperature. Decant the ether and wash the precipitate with ether. Wash ether with water, $NaHCO_3$ and water, and dry, evaporate in vacuum the ether to get the bromide (recrystallize-ethanol). 2 g of the bromide and 1.5 g piperidine (or equimolar amount DEA, etc.) in 65 ml methanol and heat in sealed tube fifteen hours at 100° (or let stand room temperature twenty-four hours). Evaporate in vacuum to get the 5-methoxy-dialkyltryptamine.

C: Reduce (II) to 5-methoxy-DET or analog with lithium aluminum hydride or diborane as described previously.

Tryptamine CA 54,13018(1960)

15.4 g indole (or substituted indole) in 40 ml ether under N_2; add to 50 ml methyl MgBr in ether (2.62 M) and reflux fifteen minutes. Cool on ice bath and add 68 ml ethyleneimine in 30 ml dry xylene. Stir 1½ hours at room temperature (add xylene to keep volume constant), reflux one hour and cool to room temperature. Add dropwise 75 ml water and take pH to 1 with HCl. Filter, wash precipitate with ether and dissolve in 150 ml hot water. Cool to 10° for fifteen minutes and decolorize. Filter, neutralize with 80 ml 10 N NaOH and stir at 5°. Filter and dry to get the tryptamine which can be converted to the dialkyltryptamine as described previously. More tryptamine can be had by adjusting pH of solution to 10 with NH_3 (or NH_4OH) and extract with methylene Cl: evaporate in vacuum.

Dialkyltryptamines BSC 1417(1965)

Dissolve 456 g o-vanillin (2-OH-3-methoxy-benzaldehyde) and 150 g NaOH in 1 L water and add dropwise 660 g benzenesulfonyl-Cl. Filter, wash precipitate with water, ethanol and recrystallize-acetic acid to get 805 g of the benzenesulfonate (I). Can probably also protect the OH group by methylating with dimethylsulfate as described elsewhere here. However, OH protection may be unnecessary to yield an active final product. Quickly add 250 g (I) to 2.5 L fuming HNO_3 at 0° and let stand five minutes. Pour onto ice and filter after melting. Wash precipitate with water, ethanol and recrystallize-acetic acid to get 171 g (II). Dissolve 170 g (II) in 2 L methanol and reflux ten minutes. With vigorous stirring add 100 g K_2CO_3 in 200 ml water and 400 ml methanol. Reflux one-half hour, filter and dissolve precipitate in boiling water; acidify to precipitate 6-NO_2-o-vanillin (about 86 g) (III). Dissolve 50 g (III) and 40 g K_2CO_3 in 1.2 L DMF and 60 ml water. Stir 1½ hours and add dropwise benzyl-Cl and continue stirring eighteen hours under N_2 if possible and in the presence of NaI. Precipitate by pouring on ice; filter, dry and recrystallize-benzene-petroleum ether to get about 52 g 2-benzyloxy-3-methoxy-6-NO_2-benzaldehyde (IV). Again, methylation can probably be done instead, and this step may be unnecessary. To 50 g (IV) in 1 L ethanol add 18 ml nitromethane, and then at -15° add 25 g K_2CO_3 (or KOH) in 40 ml water and 400 ml ethanol in small portions over a period of one hour. Keep two hours at -10⁰, then acidify with 50 ml concentrated HCl (keep temperature below ⁻0°). Add 2 L water, filter (or extract five times with ether and dry, evaporate in vacuum). Heat precipitate five minutes at 140° with 60 g anhydrous sodium acetate and 250 ml acetic anhydride and pour on ice, filter to get 52 g of the nitrostyrene (V) (recrystallize-methanol). 60 g (V) and 300 g iron filings in 1 L 80% acetic acid and heat forty minutes at 95°. Add 1 L saturated Na bisulfite and extract with benzene; dry and evaporate in vacuum (can chromatograph on alumina with benzene) to get 33 g 4-benzyloxy-5-methoxy-indole (VI) (recrystallize-benzene-petroleum ether), which can be converted to the dialkyltryptamine by any of the methods described here or as follows. 320 ml dioxane, 320 ml glacial acetic acid, 24.8 g 40% formaldehyde, 25 g DMA (or other amine) in 40% aqueous solution. To this stirred mixture at 0° add 34 g (VI) in 300 ml dioxane, let stand twelve hours and pour on ice. Basify, filter and extract with HCl; basify and extract with ether (or just dry and evaporate in vacuum after pouring on ice) to get 30 g of the gramine (VII) (recrystallize-ethyl acetate). To 20 g (VII) in 400 ml methanol add a solution of 9 g KCN in 20 ml water and 25 g methyl iodide below 10°. Stir at room temperature for twenty-four hours and evaporate in vacuum. Dissolve residue in ether, wash with dilute HCl, $NaHCO_3$, water and dry, evaporate in vacuum to get the acetonitrile (VIII) (recrystallize-ethanol). Dissolve 14 g (VIII) in 300 ml 33% ethanol solution of DMA (or other amine) and hydrogenate at 40° and 100 atmospheres pressure in presence of 3 g Raney-Ni for about three hours. Filter, evaporate in vacuum (can dissolve residue in ether, dry and evaporate in vacuum) to get 7 g of the dialkyltryptamine.

Indoleacetonitrile BER 58,2043(1925)

To the Grignard reagent prepared from 2.4 g Mg, 10 ml anisole and 16 g ethyl iodide, add 7.8 g indole in 10 ml anisole. Stir and cool to 0°. Slowly add dropwise 5.5 g Cl-acetonitrile in 40 ml anisole. Warm to room temperature and stir; heat to 60-70° on water bath (a reddish precipitate forms over about twenty minutes). Cool and purify and convert to the dialkyltryptamine as described earlier.

Tryptamines JOC 24,894(1959)

Add a 1 M excess of nitroethylene dropwise to liquid 5-benzyloxyindole (or other indole) on a steam bath over two hours. Cool, filter and recrystallize from methylene chloride-petroleum ether. Hydrogenate at two atmospheres over 10% palladium-carbon catalyst to get 5-OH-tryptamine in 30% yield. Can probably also reduce with lithium aluminum hydride or diborane methods.

4-Nitration of 5-methoxy-DMT JMC 12,321(1969)

Dissolve 35 g 5-methoxy-DMT (or analog) in 100 ml glacial acetic acid, cool to 10°, stir and add dropwise a solution of 30 ml concentrated HNO_3 in 50 ml glacial acetic acid over one-half hour. Let warm to room temperature, stir eight hours and dilute with 1 L ice-water. Filter, wash precipitate with water and dry to get 4.5 g 5-methoxy-4-NO_2-DMT (recrystallize-methanol).

Methylation of OH indoles JOC 23,1977(1958), CA 74,53525 (1971)

The procedure will not work for tryptamines. This and the two following procedures may be used to protect the OH group during subsequent steps, or simply to give the alkyloxy compounds which may be more active than the OH compounds.

Dissolve the OH-indole in ethanol (0.14 M in 50 ml) and add 28 ml dimethylsulfate and 1.2 g Na hydrosulfite. Add slowly with stirring and cooling (under N_2 if possible) 12 g NaOH dissolved in 26 ml water (keep temperature at 20-25°). Heat to 70° for one-half hour, cool and dilute with an equal volume water. Extract the yellow oil into ether-benzene and dry, filter, and evaporate in vacuum to get the methoxy-indole.

Methylation of OH-tryptamines LAC 513,16(1934)

This should work well with OH-dialkyltryptamines.

Dissolve 1 g OH-tryptamine in methanol or ether; add 2.5 g diazomethane in ether and heat twenty hours. Evaporate in vacuum to get the methoxy-tryptamine.

Acetylation of OH-tryptamines

May require modification for OH-indoles, but probably not for OH-dialkyltryptamines.

Dissolve 4.1 g OH-tryptamine in 20 ml 1 N NaOH and evaporate to dryness under N_2. Dry in vacuum at 90° and dissolve in 50 ml dimethoxyethane. Add to 1.9 g acetyl-Cl in 50 ml dimethoxyethane and stir four hours at room temperature. Add to dilute $NaHCO_3$ and $CHCl_3$; shake and dry, evaporate in vacuum the $CHCl_3$ layer to get the acetyltryptamine. This will also work with benzoyl-Cl for benzoylation and with chlorsulfonic acid for sulfonylation. Acetylation can also be done with acetic anhydride in NaOH at 5°.

Demethylation of methoxy-oxyindoles BER 96,253(1963)

The procedure may require modification for tryptamines, and indoles.

Mix 1.63 g 4-methoxy-oxyindole and 1.5 g finely ground $AlCl_3$ and heat ten minutes at 220-230° in oil bath. Cool, powder and dissolve in ice water, evaporate in vacuum to get 4-OH-oxyindole (recrystallize-water).

4-Substituted Indoles from 5-OH-indoles HCA 51,1209(1968); see also: TL 4459(1966), JOC 35,3764(1970), TET 26,3685(1970)

This method, although described for indoles, probably also works with 5-OH-tryptamine (serotonin), and 5-OH-DMT (bufotenin); *with compounds of the latter type, orally active psilocybin analogs will be obtained in one step.* Dissolve 5 g 5-OH-indole (or analog) in 25 ml ethanol. Add 5.5 g 33% aqueous dimethylamine (or other amine, e.g., piperidine) and add slowly dropwise with stirring 3.5 g 38% aqueous formaldehyde. Two minutes after the end of the addition shake with water and $CHCl_3$: dry and evaporate in vacuum the $CHCl_3$ phase to get 5 g oily 4-dimethyl-aminomethyl-5-OH-indole (I) (can chromatograph on 100 g alumina and elute with ethyl acetate). It has been claimed that this method does not work.

Alternatively (CA 72,66732(1970)), add 1.6 g 5-OH-indole, 1 g bis-dimethyl-aminomethane in 5 ml dioxane. Heat 2½ hours on water bath to get 1 g (I) (recrystallize-methanol-dimethylfor-mamide).

For the transformation of the dimethylaminomethyl substituent of (I) into methyl, aminomethyl, formyl or cyanomethyl see BSC 2046(1973).

4-Hydroxydimethyltryptamine (Psilocin) from DMT C.R. Acad. Sci. Paris 275,613(1972)

See also C.R.A.S.P. 269,51(1969) and BSC 1523(1959).

Mix 0.01 M dimethyltryptamine, 0.02 M phosphate buffer pH 7.2 containing 5 mM ascorbic acid, 0.02 M disodium EDTA and 0.01 M ferrous sulfate (CuCl may substitute) and add with stirring at 20-22° 0.02 M H_2O_2 (0.01 M may increase yield). Let reaction proceed to completion (2 hours or less) and extract with ethyl acetate. Dry and evaporate in vacuum to get about 30% yield of psilocin. The product, which contains the other OH-DMT's as well, can be chromatographed on silica thin layer with t-butanol-acetic acid-water (ACS 22,1210 (1968)) or on a 5% alumina-Nickel column or 10% alumina-Nickel plate with $CHCl_3$-methanol and the psilocin eluted with methanol.

Tryptamine from Tryptophan Synthesis 475(Sept. 1972)

Gently reflux a suspension of 250 mg L-tryptophan in 10 g warm diphenylmethane in a stream of nitrogen (if possible) for 5-20 minutes until there is no more CO_2 evolution. Cool and evaporate in vacuum or treat with 20 ml benzene saturated with dry HCl and filter, wash precipitate with hexane and dry to get about 60% yield of tryptamine.

4-Acetylation of 5-OH-indoles BCSJ 44,550(1971)

To 2 g 5-OH-indole (or analog) in 50 ml nitrobenzene add 4 g $AlCl_3$ in 50 ml nitrobenzene. Add 1.26 g Acetyl-Cl and heat three hours at 50°; evaporate in vacuum or add dilute HCl to get 1 g 4-Acetyl-5-OH-indole (recrystallize-ethyl acetate).

4-Substituted Tryptamines JMC 8,200(1965)

Heat 85.5 g 3-carbethoxy-2-piperidone and 30 g KOH in 1 L water for twelve hours at 30°. Filter, cool to 0°, add 50 ml 6N HCl. Prepare a fresh solution by diazotizing at 0-5° a mixture of 85 g 3-amino-4-Cl-acetophenone, 250 ml concentrated HCl and 750 ml water with a solution of 36 g Na nitrite in 125 ml water. Add the piperidone solution at 0° to the diazonium salt solution and stir five hours at 10°. Filter, wash precipitate with water to get 80% yield of the hydrazone (I) (recrystallize-95% ethanol). Reflux 62 g (I) in 310 ml 88% formic acid to get 40 g of the carboline (II) (recrystallize-absolute ethanol) (test for activity). Reflux 40 g (II), 100 g KOH, 480 ml ethanol and 360 ml water for eighteen hours and evaporate in vacuum. Add 480 ml water to the residue, cool and adjust pH to 6 with glacial acetic acid. Scratch glass to precipitate; filter, wash precipitate with cold water to get 41 g 4-acetyl-2-COOH-7-Cl-tryptamine (recrystallize-50% ethanol) which can be alkylated to the active dialkyltryptamine as described elsewhere here.

Alpha,alpha-DMT from Gramine or Indole-3-aldehyde JCS 7165 (1965), 3493(1958)

81 g (0.465 M) gramine (or analog), 405 ml 2-nitropropane and 20.4 g NaOH are stirred and boiled under nitrogen for 20 hours. Cool and treat with excess 2N acetic acid and ether and wash the ether layer with 2N acetic acid and water and dry, evaporate in vacuum to get ca. 90% 3-(2 methyl-2-nitropropyl)indole(Ia) (recrystallize-benzene-light petroleum). See JCS 3493(1958) for the

use of the indole-3-aldehyde to get the corresponding nitronium salt (Ib). (Ia) is then reduced (e.g., with PtO at 20°/1 Atm. in EtOH or one of the methods given elsehwere here) or (Ib) is reduced (e.g., with lithium aluminum hydride as described many times, e.g., JMC 6,378(1963)) to give the tryptamine, which can be alkylated as described. This paper also gives a route to other alpha and beta mono and disubstituted tryptamines.

Reduction of Indolyglyoxamides with Diborane JOC 38,1504 (1973), TET 1145(1968)

A solution of 7mM of the indolyglyoxamide in 160 ml tetrahydro-furan and 30 ml (30mM) 1.0M borane in THF is refluxed 2 hours and cooled. Carefully add water, then evaporate in vacuum and dissolve residue in ether. Wash twice with saturated saline, dry and evaporate in vacuum. Triturate with ether and filter to get the tryptamine-borane in ca. 30% yield. Diss. 200 mg in 2 ml xylene and 2 ml octene-1 and reflux 4 hours. Cool, dilute with hexane and filter to get ca. 80 mg white tryptamine (or dialkyltryptamine) (recrystallize-acetone-hexane etc.).

Alternatively, 63 g (0.45M) BF$_3$ etherate in 180 ml diglyme is added dropwise with stirring to 0.15 M of the indolyglyoxamide and 12 g NaBH$_4$ in 300 ml diglyme. Stir 12 hours at room temperature and evaporate in vacuum. Boil residue 2 hours with methanolic NaOH and evaporate in vacuum. Take up residue in ether, wash with water and dry, evaporate in vacuum to get the dialkyl-tryptamine in ca. 80% yield.

Indole Syntheses

4-7-Dimethoxyindoles Farmaco 25,972(1970)

Heat 1 g Cl-acetone and 7 g 2,5-dimethoxyaniline (or analog) at boiling until complete fusion; then one-half hour at 180° on a metal bath. Cool, dissolve in a little ethanol and pour into 100 ml 7% HCl. Cool, filter and dry (can distill 100/.001) to get 160 mg 4,7-dimethoxy-2-methylindole (recrystallize-benzene:petroleum ether, 1:1).

4-Carboxyindoles CJC 49,2784,2797(1971)

To a stirred solution of 0.05 M 2-methyl-5-nitroisoquino-linium iodide in 250 ml water, add a solution of 11.2 g KOH in 110 ml water and immediately add a solution of 49.4 g K ferricyanide in 300 ml water; stir one hour at room temperature. Add 50 ml methylene chloride and 100 ml benzene and stir until solids dissolve.

Separate organic phase and extract aqueous phase with benzene. Wash combined organic layers with 10% HCl, water and dry and evaporate in vacuum to get about 90% yield of 2-methyl-5-nitroisocarbostyril (I) (recrystallize-cyclohexane). Mix 8.16 g (I), 18 ml acetic anhydride, 5 drops concentrated sulfuric acid and heat on steam bath four hours. Cool to room temperature, add 25 ml water and heat on steam bath until a single phase is produced (about ten minutes). Pour into a large volume of water, extract with methylene chloride, wash methylene chloride five times with water and dry and evaporate in vacuum to get 6.8 g 2-methyl-4-acetyl-5-nitro-isocarbostyril (II) (recrystallize-benzene). 1.64 g (II), 200 mg 10% palladium-carbon in 100 ml ethanol and hydrogenate at room temperature and initial pressure of 60 pounds per square inch for one hour. Transfer to 500 ml round bottom flask under N_2. Evaporate in vacuum and add solution of 8.4 g KOH in 120 ml 50% dimethylsulfoxide to residue. Stir and reflux seventeen hours under N_2 and then add concentrated HCl until the dissolved silica precipitates from the basic solution. Filter, wash precipitate with water and acidify filtrate with concentrated HC1. Extract filtrate with ethyl acetate and then shake ethyl acetate with Na carbonate (100g/L). Acidify solution with concentrated HCl and extract with ethyl acetate. Wash extract with water and dry, evaporate in vacuum to get 400 mg 2-methyl-indole-4-COOH (recrystallize-60% ethanol). Can dissolve in methanol and chromatograph on 5 g silica gel-250 ml ethyl acetate elutes the indole.

4-OH-Indole JCS 1605(1948)

Dissolve 0.7 g KOH in 10 ml ethanol, cool to -10° and add with stirring over twenty minutes to 1 g 2-nitro-6-OH-benzaldehyde and 0.4 g nitromethane in ethanol at -10°. Acidify with cold HCl, add water and extract with ether. Dry and evaporate in vacuum the ether to get a yellow oil. Warm the oil with 2 g Na acetate and 3 ml acetic anhydride for fifteen minutes. Add 20 ml water, separate the precipitate and add water to crystallize about 1.3 g 2,beta-dinitro-6-acetoxy-styrene (I) (recrystallize-ethanol). Warm gently a solution of 0.5 g (I), 2 g iron filings, 4 ml ethanol and 4 ml glacial acetic acid until hydrogen evolution is vigorous and continue to warm gently for ten minutes. Filter, wash precipitate with a little warm ethanol and add water to filtrate. Basify with Na carbonate, extract three times with ether and dry and evaporate in vacuum the ether to get about 0.1 g 4-acetoxyindole (II) (recrystallize-petroleum ether). If 4-OH-indole is desired, dissolve 0.1 g (II) in 3 ml methanol and saturate at 0° with NH_3 and keep below 5° for sixteen hours. Evaporate in vacuum, add 50 ml water and extract with ether, dry and evaporate in vacuum to get 4-OH-indole (recrystallize-petroleum ether).

4-Substituted Indoles MON 100,1599(1969), 101,161(1970)

To a solution of 1.84 g Na metal in 60 ml ethanol at 5-10° add, over ½ hour with vigorous stirring a mixture of 0.08 M ethyl azidoacetate and 0.02 M 2 (or 2,5; 2,3 etc. but not 6) substituted benzaldehyde and continue stirring at 5-10° until nitrogen evolution ceases (about ½-1 hour); then stop immediately and rapidly evaporate in vacuum ½ the ethanol (keep temperature below 30°). Basify the solution with solid NH_4Cl, dilute with 500 ml water and extract 3 times with ether. Filter, wash with water to neutrality and dry, evaporate in vacuum the ether (or can dissolve the residue in petroleum ether, or 1:1 petroleum ether:benzene for methoxy compounds, and filter through silica gel) to get the ethyl-alpha-azidocinnamates (I) in about 50% yield. Store in freezer until used in next step. Dissolve 1 g (I) in 100 ml p-xylol and reflux 10 minutes. Evaporate in vacuum (or add 5 ml pentane, filter, evaporate in vacuum) to get about 90% yield of the 4 substituted-2-carbethoxyindole which can be decarboxylated as described elsewhere here.

4-Methoxy-indole JCS 3909(1952)

4.5 g 2-methoxy-6-nitro-benzyl-Cl (JOC 6,217(1941)), 4.5 ml ethanol, 1.6 g KCN, 1.5 ml water; reflux six hours, cool and extract with ether. Wash with water, dry and evaporate in vacuum the extract to get 4 g of the cyanide (I). 4.2 g (I), 125 ml concentrated HCl; reflux six hours, cool and filter. Add precipitate to dilute aqueous Na carbonate; stir and warm until dissolved and filter. Acidify and let stand one hour at 0°; filter to get 1.8 g 2-methoxy-6-nitro-phenylacetic acid (II). 5 g (II) in 100 ml glacial acetic acid catalyzed with palladium-carbon at room temperature gives 3.8 g 4-methoxy-indole. Filter, evaporate in vacuum and recrystallize-1:5 ethanol/water.

Indoles CA 74,87819(1971)

To 30 g polyphosphoric acid (prepared by adding 2:1 P_2O_5:85% phosphoric acid) add 3.3 g phenylhydrazine and 4 g phenylacetone (or equimolar amount ethylacetone) and heat twenty minutes at 130-140° in N_2 stream. Pour over 200 ml water and extract with methylene chloride to get about 45% yield 2-benzylindole or 30% 2-

ethyl-indole (recrystallize-ether).

Indole-3-acetic acid CA 72,66815(1970)

To 21.6 g phenylhydrazine dissolved in 300 ml 0.3N sulfuric acid, add 9.8 g concentrated sulfuric acid. Add dropwise with stirring and heating at 100°, 11.6 g methyl-beta-formyl-propionate in 300 ml 0.3 N sulfuric acid and continue heating six hours to get 14 g indole-3-acetic acid. Convert to the dialkyltryptamine as already described.

α, α -Dimethyl-indoleacetic acid ACS 8,122(1954)

117 g indole in 1 L acetone; add 220 g NaOH pellets and stir in three-neck, 3 L flask with thermometer and dropping funnel. Cool in ice bath and add CHCl₃ dropwise over about two hours, keeping temperature below 15°. Let warm to room temperature over one hour with stirring and continue stirring four hours. Warm on water bath and evaporate the acetone. Dissolve the residue in 1.5 L water and extract four times with ether (dry and evaporate in vacuum the extract to get about 50 g unreacted indole). Acidify the aqueous phase with dilute sulfuric acid and extract the oil with ether. Wash with water and dry and evaporate in vacuum the extract to get the title compound (recrystallize-aqueous methanol), which can be converted to the dialkyltryptamine as described elsewhere here.

4-Nitroindole JACS 80,4621(1958) -cf. Chem. Het. Cpds. (Russian) 37(1973).

21 g ethyl-pyruvate-meta-nitro-phenylhydrazone (I) in 100 g polyphosphoric acid; warm in oil bath with stirring and cool to keep temperature below 125°. Stir at 105-115° for one-half hour and then add ice water. Filter, extract water with ether to remove unreacted (I), wash precipitate with water and dry to get 9.5 g ethyl-4-nitroindole-2-carboxylate (II) (recrystallize-ethanol or 800 ml benzene). It may be possible to use this directly for diakyltryptamine synthesis, otherwise proceed as follows. Dissolve 9.7 g (II) in 75 ml hot ethanol and add solution of 7.5 g KOH in 18 ml water, let stand six hours at room temperature. Add 350 ml warm water and add the mixture to excess dilute HCl so that the pH is acid. Filter, wash precipitate with water and dry to get 9 g 4-nitroindole-2-COOH (III). 11.3 g (III) and 1.1 g CuO or Cu powder in 90 ml quinoline and reflux 2 hours. Add 100 ml concentrated HCl and 200 g ice; filter and extract both precipitate and filtrate with ether. Evaporate in vacuum or wash with dilute NaHCO₃, water and

dry and evaporate in vacuum to get the title compound (recrystallize-dilute ethanol).

Methoxyindoles from Methoxynitrostyrenes JOC 22,331(1957)

Illustrated for 5,6,7-trimethoxyindole (for the trimethyl-indole see JOC 25,1542(1960)). 2-methoxy-beta-nitro-styrene, if nitrated at the 6 position and used in place of (I), should give 4-methoxy-indole.

7.9 g 3,4,5-trimethoxy-beta-nitrostyrene (preparation described in mescaline section) in 40 ml acetic anhydride at -8°: stir well and add dropwise 5 ml fuming HNO_3. Stir twenty minutes and pour on 200 ml ice water. Add some solid Na carbonate, filter, wash with water to get the 2-nitro derivative (I) (recrystallize-aqueous ethanol). 2.5 g (I) in 18 ml ethanol; add 8.8 g iron powder and 18 ml glacial acetic acid and stir and warm (cooling is eventually necessary). Stir about ten minutes until solution thickens and about five minutes more, then add 50 g Na bisulfite in 220 ml water and extract the indole with 5X 200 ml ether. Wash combined extracts with $NaHCO_3$, water, and dry and evaporate in vacuum to get the indole (can purify on alumina, eluting with 1:1 benzene:petroleum ether, discarding the precipitate that won't dissolve in the benzene-petroleum ether).

4-Substituted Indoles from Pyrrolyl Aldehydes J. Prakt. Chem. 315,295(1973)

A mixture of 9.5 g pyrrolyl-2-aldehyde, 29.2 g dimethyl-succinate and NaH (9.6 g of 50% suspension in oil) in 100 ml benzene is stirred at room temperature 6 hours, cooled and carefully acidified with glacial acetic acid. Add water and ether and dry, evaporate in vacuum or work up (JACS 72,501(1950), JCS 1025(1959)) to get ca. 17 g (80%) 3-methoxycarbonyl-4-(2'-pyrrolyl)-3-butenoic acid (I) (recrystallize-acetone-benzene). A mixture of 12 g (I), 7 g sodium acetate and 70 ml acetic anhydride is left overnight at room temperature with occasional shaking. Then gradually raise the temperature to 70-75° over 2 hours, maintain for 4 hours and work up (see JCS 1714(1955), 986(1958)) to get ca. 8 g (60%) methyl-4-acetoxy-indole-6-carboxylate (II) (recrystallize-petroleum ether). If desired, this can be converted to 4-OH-indole-6-COOH and 4-methoxyindole-COOH as described in the ref. or decarboxylated as described elsewhere here. If the 1-methyl cpd. is used, 1-Me-indole results.

Dialkyltryptamines from Indoles BSC 1424(1973)

Dissolve 2.75 g anhydrous sodium carbonate in a mixture of 50 ml acetic acid and 25 ml propionic acid. Add 11.7 g (0.1M) indole and 8.8 g (0.11M) dimethylamine hydrochloride (or other amine) in 5 ml diglyme. Cool to between zero and minus 5° and add 11 g chloracetaldehyde hydrate (containing 75% chloracetaldehyde-0.1M). Stir 1 hour, keeping temperature below zero and put in refrigerator at ca. 5° for ca. 5 days. Pour on crushed ice and filter out the white, flocculent ppt. Shake the cold acid filtrate with cold ether (vigorous cooling) and separate the ether. Slowly with stirring neutralize by adding NaOH dropwise. Rapidly decant the ether phase, dry rapidly over sodium sulfate and quickly, with stirring, at a low temperature add HCl in ether. Decant the ether and treat the thick oil which formed with a small amount of absolute ethanol to bring about crystallization (add dry ether if necessary). Can recrystallize 3 times from ethanol-ether or acetonitrile to get 8.2 g (32%) alpha-chloromethylgramine (I).

Add 1 g (I) to a suspension of 2.5 g $NaBH_4$ in 75 ml diglyme heated to 55° with stirring. Heat to 85-90° and keep at this temperature for 48 hours (the yellow color disappears). After 24 hours, cool and pour on ice. Extract with ethyl acetate and evaporate to get about 37% yield of the oil DMT (or analog). Can purify on 1:1 silica:celite column; benzene elutes the skatole, methanol the DMT (recrystallize-hexane).

4-Acetylindoles from 5-Hydroxyindoles BCS 2046(1973)

Heat 1.1 g 2-ethoxycarbonyl-5-hydroxyindole(5-OH-indole and perhaps 5-OH-DMT will work) and 0.9 g powdered $AlCl_3$ at 145° in orthodichlorobenzene. Add 2-3 ml acetylchloride and keep the temperature at 145-150° for 2 hours. Cool and hydrolyze with ice cold 2N HCl. Extract with chloroform and dry, evaporate in vacuum to get 4-acetyl-2-ethoxy-carbonyl-5-hydroxyindole in ca. 50% yield. This paper also gives a method for borminating the starting cpd. in the 4 position.

5-Methoxy-indole REC 81,317(1961)

Again, if 3,5-dimethoxyaniline is used, 4,6-dimethoxy-indole should result.

To 123 g p-anisidine (p-methoxy-aniline) in 1 L 15% HCl; add 1 Kg ice and while stirring, quickly add dropwise below the surface, a solution of 75 g $NaNO_2$ in 200 ml water (temperature below 5°). Add 5 g activated carbon, stir and filter and quickly add filtrate with stirring to a mix of 160 g ethyl-alpha-methyl-acetoacetate in 1 L methanol, 1 Kg ice and 820 g Na acetate. Stir two hours and extract with 4X500 ml benzene. Wash extract with water, dry and evaporate in vacuum and dissolve residue in ½ L ethanol. Cool to 0° and quickly add ½ L cold ethanol saturated with dry HCl gas (violent boiling). Stir one hour, add 100 ml hot water and refrigerate twelve hours. Filter, wash precipitate with 2X100 ml ethanol, 2X200 ml hot water and dry to get 135 g ethyl-5-methoxy-indole-2-COOH (I). (I) might be used directly to synthesize a dialkyl-tryptamine. 135 g (I) and 56 g KOH in 1 L 90% ethanol; reflux 1½ hours and pour into 6 L water. Acidify with concentrated HCl, filter, wash with water and dry to get 108 g 5-methoxy-indole-2-COOH (II). To a stirred mixture of 45.6 g (II) in 60 ml fresh quinoline, add 2 g Cu chromite and heat to 210-230° in a metal bath until CO_2 evolution stops (or just heat at 210-230° five minutes under N_2). Cool, filter, dry and evaporate in vacuum or add ½ L ether, filter and extract filtrate with 3X200 ml 2N HCl, 3X200 ml water, 2X100 ml 2N NaOH, 2X100 ml water and dry, filter, evaporate in vacuum to get 24 g 5-methoxy-indole.

5-Nitroindole JGC 29,2508(1959); see also JOC 20,1541(1955)

11.9 g indoline dissolved in 150 ml acetic anhydride. Add 8.5 ml concentrated HNO_3 dropwise with stirring (temperature 10-12°). Pour the mixture into water and filter. Add the precipitate (or 1-acetyl-indoline, prepared by boiling indoline in acetic anhydride) to 100 ml concentrated HCl and boil one-half hour. Basify with $NaHCO_3$ (can filter to remove black precipitate formed after adding first portion of $NaHCO_3$) to get 12 g 5-nitro-indoline (I) (recrystallize-heptane). 7.9 g (I) and 11.7 g chloranil in xylene. Reflux ten hours, cool, evaporate in vacuum or wash two times with 20% NaOH, filter, wash xylene with water, 50% HCl, water and dry, evaporate in vacuum to get 5.4 g 5-nitro indole (II). Can reduce (II) to the aminoindole as described for preparation of 4-amino-DMT or as follows. To 5.4 g (II) in 50 ml ethanol add a little Raney-Ni and add dropwise over two hours 100 ml hydrazine hydrate to the boiling solution. Filter, evaporate in vacuum (can distill 190/6) to get 5-NH_2-indole (recrystallize-heptane). Can benzoylate by dissolving in benzene and shaking with 20% NaOH and excess benzoyl-Cl, let stand twelve hours and dry, evaporate in vacuum (recrystallize-benzene). The dialkyltryp-tamine of the benzoylated compound may be more active than that of the amino compound.

4-CN-indole JACS 71,761(1949) (See BER 89,270 (1956), 90,1980(1957) for a more involved method.)

To 23 g Na in 350 ml ethanol add 146 g ethyl-oxalate and 171 g 2-nitro-6-Cl-toluene and reflux forty minutes. Dilute the red solution with water and steam distill until no more starting material is distilled. The aqueous residue is filtered, acidified with HCl and filtered to get 102 g 2-nitro-6-Cl-phenylpyruvic acid (I) (recrystallize-benzene). Add 81 g (I) in dilute NH_4OH to a solution of 560 g $FeSO_4.7H_2O$ and 230 ml concentrated NH_4OH and 2 L water and boil five minutes. Filter, wash precipitate with dilute NH_4OH, water and acidify filtrate with dilute HCl to get 60 g 4-Cl-2-indole-COOH (II) (recrystallize-aqueous ethanol). 9.78 g (II) and 6.7 g CuCN in 35 g quinoline and reflux (about 237°) for twenty hours. Pour the hot solution into a mixture of 25 ml concentrated HCl and ice. Stir and filter; wash precipitate with water and extract the filtrate and precipitate three times with ether. Wash the ether with HCl, water and dry, evaporate in vacuum to get 3.6 g 4-CN-indole (recrystallize-water). Or, heat (II) alone at 290° until fusion; then heat at 250° for ten minutes until CO_2 evolution ceases to get 4-CN-indole. For conversion to 4-formyl-indole see HCA 51,1616(1968).

4-Alkyloxyindoles TET 24,6093(1968), JMC 13,983(1970) and a longer route in CT 274(1970).

Add solution of 150 g 2-amino-6-nitrotoluene (prepared by reduction of 2,6-dinitrotoluene with H_2S in ammoniacal ethanol and recrystallize-water or ethanol) in 550 g concentrated sulfuric acid slowly to 5 Kg crushed ice and then add dropwise a solution of 75 g Na nitrite in 100 ml water with stirring and cool (0-5°). Continue stirring two hours and add 2 L 10% sulfuric acid containing 0.1% Cu sulfate. Heat at 75-80° twelve hours or until N_2 evolution ceases. Filter, cool to precipitate 135 g 2-OH-6-nitrotoluene (I) (re-crystallize-aqueous ethanol) (can also prepare by nitrating o-cresol). Add 20 g (I) in 100 methanol to a solution of 5.8 g Na metal in 60 ml methanol or ethanol. Add dropwise with stirring, 34 g diethyl or dimethylsulfate. After addition reflux one hour, evaporate in vacuum most of the methanol, add 100 ml water and extract with 3X50 ml ether. Wash ether with 50 ml 5% NaOH, 2X50 ml water: dry, filter, evaporate in vacuum (can distill 116/2.2 for ethyl-, 94-6/0.5 for methyl-) to get 20 g 2-alkyloxy-6-nitrotoluene (II). Can also use 1-Br-propane and reflux four hours; evaporate in vacuum the ether extract and distill residue (180/17) to get the 2-n-propoxy-6-nitrotoluene (recrystallize-aqueous methanol). To a suspension prepared from 7.8 g K metal (Na may do) and 25 ml absolute ethanol in 160 ml dry ether, add dropwise a solution of 0.1 M (II) and 29.2 g diethyloxalate in 50 ml dry xylene. Stir four hours and let stand at room temperature three days. Extract the precipitate with 100 ml ice water and extract the resulting dark red aqueous layer with 3X50 ml ether. Filter and remove the residual ether by blowing air through the aqueous solution. Cool in ice bath and treat alternately with a little 20% NaOH and 30% H_2O_2 (16 ml each total) until the dark red color is gone. Filter and acidify with concentrated HCl to get 9 g of the 2-alkyloxy-6-nitrophenylacetic acid (III). 2 g (III) in 40 ml glacial acetic acid. Add 0.1 g 10% palladium-carbon and hydrogenate at room temperature and atmospheric pressure about one hour. Filter, evaporate in vacuum to get 1.2 g 4-alkyloxyindole (recrystallize-toluene). For the n-propoxy compound, hydrogenate one-half hour in 30 ml 5% NaOH, filter, acidify with concentrated HCl and heat on steam bath for one-half hour. Cool and filter and recrystallize-aqueous ethanol. If desired, these indoles can be dealkylated as described elsewhere here.

Alternatively, dissolve 45 g (III) in 600 ml water with the minimum amount of 2N NaOH (about 80 ml). Add $Na_2S_2O_4$ to the well stirred solution in small amounts until no further increase in temperature occurs. Add 2N NaOH dropwise, at the same time add further Na hydrosulfite, until the red color disappears (about 80 ml NaOH, 70 g Na hydrosulfite). This takes about one hour and final temperature is about 35°. Stir in 200 ml dilute (1:1) HCl to precipitate the alkyloxyindole-COOH (IV) which can be used as is, or decarboxylated with quinoline and Cu powder by heating to 245° and cool, filter and evaporate in vacuum as described elsewhere here.

Methoxyindoles from Methoxyanilines JOC 20,1454(1955), 23,19 (1958)

Illustrated for preparation of 4,5,6-trimethoxyindole (IV). 1.8 g 3,4,5-trimethoxyaniline in 10 ml glacial acetic acid; add 1.9 g ethyl-oxomalonate-dihydrate and heat ten minutes on steam bath. Let stand two hours at room temperature and add 125 ml water and then solid ammonium carbonate till pH is 8 to precipitate 2.7 g 4,5,6-trimethoxy-3-OH-3-carbethoxyindole (I) (recrystallize-ben-zene-petroleum ether). 4.5 g (I) in 50 ml 5% NaOH and heat on water bath with aeration ten minutes. Take pH to 4 with dropwise addition of formic acid. Filter, wash with water and recrystallize from aqueous ethanol to get 1 g 4,5,6-trimethoxyisatin (II). To a stirred suspension of 7.2 g (II) in 30 ml boiling water, add 7 g Na hydrosulfite and let clear solution stand twelve hours in refrigerator.

Filter, recrystallize the precipitate from boiling water and dry to get the dioxindole (III). To a stirred slurry of 1.9 g lithium aluminum hydride in 100 ml dry ether add 4.3 g (III) in 120 ml dry benzene and reflux four hours. Cool and carefully add a little water. When bubbling stops, filter and dry, evaporate in vacuum the organic layer to get (IV).

4-Methoxyindole BSC 1335(1966)

Methylate p-Cl-phenol to get p-Cl-methoxy-benzene which is then converted (BSC 643(1953)) to 5-Cl-2-methoxy-benzyl-Cl (I). 50 g (I) in 50 ml dimethylformamide is added to a solution of 16 g KCN in 80 ml dimethylformamide and 70 ml water at 70°. Add a pinch of KI and heat at 85-90° 1½ hours. Pour on cold water; filter, dry (can purify by distilling 167/12) to get 38 g 5-Cl-2-methoxy-phenyl-acetonitrile (II). Alternatively, dissolve 29 g (I) in 200 ml acetone, add 8 g NaCN and 1.6 g NaI and reflux twenty-four hours. Cool, filter, and evaporate in vacuum to get 21 g (II). 174 g (II) in 174 g concentrated sulfuric acid, 174 ml water and 174 g glacial acetic acid; reflux and stir three hours. Cool, filter and dissolve the brown crystals in Na carbonate solution (can add decolorizing carbon and filter) and acidify to precipitate 179 g (III) (recrystallize-ethanol) (2-methoxy-5-Cl-phenylacetic acid). Alternatively, dissolve 50 g (II) in 350 ml methanol containing 15% NH_3 and hydrogenate at 70 Kg pressure, 50° in presence of Raney-Ni (or other reducing method) to get 40 g colorless liquid (after filtering, and evaporating in vacuum) 2-methoxy-5-Cl-β-phenethylamine (VIb). 76 g (III) carefully mixed with 60 g $SOCl_2$ and heated until dissolved. Let stand twelve hours and evaporate in vacuum to get 58 g yellow oily 2-methoxy-5-Cl-phenylacetyl-Cl (IV) (can distill 145/10). See preparation of mescaline for alternatives to this and other steps. 6 g (IV) and 14 ml 10% NaOH are added at the same time to 10 ml 33% aqueous methylamine at 0° and stirring is continued for one-half hour. Filter, wash precipitate with water and dry to get 5 g of the N-methyl-phenylacetamide (V). Carefully add 20 g pure and dry (V) to 10 g lithium aluminum hydride in 800 ml dry ether (can first reflux lithium aluminum hydride and ether twelve hours under N_2 and filter), reflux about four days and carefully add a little water until no more bubbles. Filter, extract filtrate with 2N HCl and extract the HCl with ether (evaporate in vacuum the ether to recover 3 g (V)). Basify the HCl extract and extract it with ether. Dry and evaporate in vacuum the ether to get 8 g oily N-methyl-2-methoxy-5-Cl-beta-phenethyl-amine (VIa) (can distill 110/0.5). 12 g (VIa or b) and 4.5 g DEA or

DMA in 500 ml ether; mix rapidly with 270 ml 0.9 M phenyl-Li, boil fifteen hours and extract as for (VI) or as described previously to get 8 g oily 4-methoxy-indoline (or its 1-methyl derivative) (VII). Alternatively, add 36 g naphthalene to 300 ml tetrahydrofuran and add 11 g Na metal cut in small pieces. Reflux and stir three hours and add 18 g (VI) and 8 g DEA in 200 ml tetrahydrofuran rapidly and boil twelve hours. Evaporate in vacuum, dissolve the oily residue in 2N HCl and extract with ether. Proceed as described to get (VII). 4 g (VII) in 200 ml dry pyridine; add to 6 g Cu chloride in 400 ml pyridine and reflux 1½ hours. Pour on water and extract with ether. Wash extract with 4N HCl and then water and dry and evaporate in vacuum the ether to get 2 g of the indole (VIII). Alternatively, dissolve 4 g (VII) and 9.5 g cinnamic acid in 700 ml mesitylene, add 1 g 5% palladium-carbon and reflux five hours. Filter, wash with HCl and $NaHCO_3$ and dry and evaporate in vacuum the mesitylene to get the red, oily (VIII) (can chromatograph on alumina and elute with benzene-petroleum ether).

5-OH-indole JCS 2525(1952)

To a solution of 4.3 g 2,5-dihydroxyphenylalanine and 2 g $NaHCO_3$ in 150 ml water, add with stirring during ten minutes, a solution of 13 g K ferricyanide and 3 g $NaHCO_3$ in 200 ml water (dark solution turns pale yellow). Extract with 3X200 ml ether and dry, evaporate in vacuum to get 2.3 g 5-OH-indole.

4-Indolecarboxylic Acid CPB 20,2123(1972)

Heat 3-nitrophthalic anhydride with ammonium carbonate to get 3-nitrophthalimide (I). Dissolve 4.3 g (I) in 50 ml 90% methanol and add 1.9 g sodium borohydride over 30 minutes while stirring vigorously at room temperature. Stir 2 hours, acidify with 20% HCl, evaporate in vacuum and treat the dry residue with acetone. Evaporate in vacuum to get 3.9 g (88%) 3-OH-4-nitrophthal-imidine (II) (recrystallize from acetone). Dissolve 3.9 g (II) in 40 ml 20% HCl and stir for 10 hours on water bath at 80-90°. Distill off HCl and stir residue with acetone. Filter and evaporate in vacuum to get 3.4 g 3-OH-4-nitrophthalide (III) (recrystallize from $CHC!_3$ and can purify on column). Prepare an ether solution of CH_2N_2 and add to 1.93 g (III) in a 100 ml flask until a reaction is no longer evident. Add acetic acid to decompose excess diazomethane and evaporate in vacuum to get about 2 g of 2-methoxycarbonyl-6-nitrostyrene oxide (IV) (can purify on column). Dissolve 560 mg (IV) in 50 ml absolute methanol, add 50 mg PtO_2 and hydrogenate as described elsewhere here (other reducing methods should work). Filter, evaporate in vacuum and recrystallize from benzene to get 270 mg methyl-4-indolecarboxylate(V). Dissolve 250 mg (V) in 2 ml 0.05M KOH and stir at room temperature 6 hours. Neutralize with 10% HCl carefully to prevent excessive heat and collect the crystals by filtration. Dry to get 126 mg 4-indolecarboxylic acid.

Indole Russian Patent 306,126(29July1971)

To a stirred suspension of 1.9 g orthocarbamoyl cinnamamide in 50 ml methanol add 26 ml 0.77 N NaOCl and heat in a distillation apparatus at 40° for 2 hours, or until no more indole is distilled off (can use the indole tests described earlier). Extract the distillate with $CHCl_3$ and dry, evaporate in vacuum (or steam distill the solvent) to get about 45% indole.

Indole German Patent 2,052,678(6May1971)

A molar ration of 2-(o-nitrophenyl)-ethanol to reducing gas of 1:5 is best (greater than 90% yields). The catalyst is Al_2O_3 or silica gel containing 7-14% by weight of copper (or Ni, Co, Cu chromite, etc.), with some potassium sulfate if necessary. Reducing gas is NH_3 or H_2, which can be mixed with nitrogen. The temperature is 250-300°. The 2-(o-nitrophenyl)-ethanol is vaporized over 30 ml catalyst in a quartz tube 56 cm long with an inner diameter of 15 mm, containing along its whole length a thermoelement tube with an outer diameter of 8 mm. The tube is filled to 28 cm with quartz glass fragments. The gas flow rate is 80-240 ml/minute with a contact time of 7-14 seconds. Compare CA 79,105065-66(1973).

Indole Analogs

Replacement of one or more of the C or N atoms of the indole nucleus by atoms of N, C, O or S will result in dialkyltryptamine analogs some of which are very likely to be psychedelic. Among the indole analogs are isoindole, indene, indazole, diazindoles, benzofuran, benzothiophene and benzimidazole. The syntheses of several such compounds follows.

Benzimidazole Analog of DET JCS 1671(1957)

17.3 g benzimidazole dissolved in 300 ml ethanol containing 3.4 g Na metal. Slowly add 22.4 g alpha-Cl-N, N-diethyl-acetamide (or 18.2 g dimethyl analog) in 100 ml ethanol and reflux four hours. Filter, evaporate in vacuum to get 2-1'-benzimidazolyl-N,N-diethyl-acetamide (I) (recrystallize-acetone). Add 16 g (I) to 6 g lithium aluminum hydride in 300 ml tetrahydrofuran and reflux six hours. Cautiously add a little water and filter, evaporate in vacuum to get the DET analog (recrystallize-water). (I) may be active.

Alternatively, dissolve 10 g N,N-diethyl-N′—o-nitrophenyl-ethyl-enediamine in 100 ml ethanol. Add Raney-Ni and 5 atmospheres hydrogen and hydrogenate one hour. Filter, evaporate in vacuum and reflux the oil forty minutes with 100 ml 4N HCl and 20 ml 87% formic acid. Basify with NH_4OH; extract four times with $CHCl_3$ and evaporate in vacuum to get the DET analog (recrystallize-water).

Indene Analogs of DET JMC 10,856(1967), JOC 30,3231(1965).

Illustrated for 6-methoxy-DET analog.

Prepare 6-methoxy-1-indanone (I) (JCS 1986(1962)) using polyphosphoric acid made by diluting 500 g of the commercial acid with 120 g 85% phosphoric acid. 2.5 g (I) in 176 ml ether and reflux one hour with 0.27 g lithium aluminum hydride. Cool and carefully add water and filter when bubbling stops (can use Celite filter aid). Dry and evaporate in vacuum and store twelve hours at -15° (under N_2 if possible) to precipitate the white 6-methoxy-1-indanol (II) (recrystallize-n-hexane). 2.5 g (II) in 73 ml benzene and reflux one-half hour with 0.2 g p-toluenesulfonic acid. Cool, add water and separate the phases. Extract the aqueous phase with ether and combine with benzene phase and dry, evaporate in vacuum to get 5-methoxy-indene (III) (can distill 110-45/10). 1.53 g (III) and 1.39 g N,N-diethyl-aminoethyl-Cl.HCl in benzene (prepare the free base in benzene as described previously). Reflux four hours with 0.42 g sodamide, cool, wash with water and dry, evaporate in vacuum to get the indene analog of 6-methoxy DET as a dark liquid (can crystallize as oxalate). Alternatively, dissolve 2.51 g (III) in ether and treat (under N if possible) with 12 ml 1.6M buty-Li in hexane at 0-10°. After two hours cool to -30° and add 12 ml more of butyl-Li. Add ether suspension of 2.5 g N,N-diethylaminoethyl-Cl. HCl over one-half hour and warm to room temperature. Filter, evaporate in vacuum to get the 6-methoxy-DET analog.

The following are some other syntheses of indole analogs:

Indene JOC 37,1545(1972); TL 2869(1972); CT 7,205(1972)

Indazole analog of 5-benzyloxy-DMT JACS 79,5246(1957), 80,966(1958)

Benziminazole analog of 5-OH-tryptamine JCS 1671(1957)

5-BR-benzothiophene analog of DMT JMC 10,270(1967)

Tryptamine analogs using pyrrolidine and piperidine JMC 10,1015 (1967)

Benzofuran derivatives LAC 662,147(1963), CA 60,2901(1963)

Benzothiophene preparation JOC 10,381(1945)

4-Azindoles CA 72,66934(1970), 73,56000(1970)

Isoindoles J.Prakt. Chem. 312,440(1970)

Various analogs JMC 9,819(1966), 10,856(1967), 13,1205 (1970); ACS 17,2724(1963); JACS 77,4324(1955); J.Het. Chem. 6,775(1969); JCS (C) 2317(1959), 1612(1969), 498 (1970); JOC 37,51(1972). CT 7,205(1972)

The following are syntheses of indoles and derivatives: For an extensive review of indole syntheses and reactions see CHEM-ISTRY OF HETEROCYCLIC COMPOUNDS (W.J. Houlihan, Ed.), 25, parts 1, 2 and 3 (1972-3).

5-OH-DMT (which can be substituted in the 4 position as described here) from 2,5-dimethoxybenzaldehyde JCS 1165(1954)

5-Nitro-DMT (in six steps from tetrahydrofurfuryl alcohol) JACS 75,1880(1953)

N-acylation of tryptamines JPS 58,563(1969), cf. JACS 74,101 (1952)

4-Acetylindole-3-acetic acid (in ten steps from m-amino-benzyl alcohol) BER 87,229(1954)

4,5-Benzindole JOC 24,565(1959)

5-methyl-indole JOC 24,2030(1959)

Indolyl-3-acetaldehyde JCS 3172(1952)

DMT derivatives JACS 77,4322(1955)

5-substituted indoles JCS 1424(1965), Israel J. Chem. 5,129 (1967)

5-6-Methylenedioxyindole JCS 78(1949)

Diphenyltryptamine CA 73,109616(1970)

Tryptamines (via ethyleneimine) CA 55,19897(1961)

Indoles (from phenylhydrazones) CA 72,66814(1970)

Indoles (possible route to methylethyltryptamine, using N-ethyl-aziridinium tetraflouroborates) Angew. Chem. 79,188 (1967)

3-(4-pyridyl)indoles (if this is not active, the 4-n-alkyl-pyridyl analog will probably be) J. Het. Chem. 7,1071(1970), ACS 22,1064 (1968)

Indoles (from o-acylanilines and dimethylsulfonium methylide) GCI 100,652(1970), TL 679(1969)

Indoles BSC 643,741(1960), 1051,2263(1962), 861,2175 (1965), 3359(1966)

Indole Grignard Reagents (review) Adv. Het. Chem. 10,43(1969)

Indoles (reviews) Het. Compounds (Elderfield-Ed.) 3,1(1952); Prog. Drug Res. 6,75(1963); THE INDOLE ALKALOIDS, W. Taylor (1966); THE CHEMISTRY OF INDOLES, R. Sundberg (1970)

4-Aminoindole from 5-Br-indole C.R. Acad. Sci. Paris 265,110 (1967)

Indole derivatives JPS 60,304(1971); AP 304,73(1971)

5-OH-indoles from substituted phenylethylamines TL 723(1970)

4-Alkylindoles from 4-Br or iodoindoles BER 104,2027(1971)

Indoles from cyclohexanones and allylamines C.R. Acad. Sci. Paris 272,1509(1971)

For a general, simple high yield indole synthesis from anilines and methylthioacetaldehyde etc. see JACS 95,588,591,2718,6508 (1973). For indoles from N-(β -hydroxy-ethyl aniline esters see BSC 2485(1973). For a 2-acyl-indoles in one step from orthoamino-ketones and alpha-haloketones or 2-carboxyindoles from sulfo-namides of ortho-aminocarbonyls see JOC 38,3622-24(1972). Indole and 5-Br-indole in 4 steps from beta-naphthol see Chem. Het. Cpds. (Russ.) 753(1973). Indole-JOC 37,3622(1972).

4-substituted indoles: from N-methylpyrrole in 5 steps Compt. Rend. 276C,1327(1973); in 20% yield by irradiation of 1-substituted indoles TL 2451(1973); in one step Swiss Patent 536,840(29 June 1973); French Patent 2,154,485(15 June, 1973); from para-benzoquinone etc. German Patent 2,145,573(15 March 1973); from 4-aminobenzofurans CA 78,147795(1973) (Japanese Patent 73 08,777(03 Feb 1973)).

Indoles Review: Int. Rev. Sci. Orgn. Chem. Ser.1 4,29(1973). Sulfur analog of psilocin JCS (P.T.1) 3011(1972); JHC 10,297 (1973); DMT analog CPB 19,603(1972). Benzothiophenes from thiophenes JOC 38,1056(1973).

Dialkyltryptamines from tryptamines: CA 78,147730(1973).

Tryptamines: J. Het. Chem. (Russ.)213(1973).

Indoles from Nitrosoanilines JHC 10,883(1973); from N-(beta-hydroxyethyl)aniline by thermal cracking: BSC 2485(1973); from isonitrile and diazomethane: TL 2133(1973).

CA 79,146321(1973) gives a one step synthesis of tryptamine from phenylhydrazine and 4-chlorobutyraldehyde by refluxing in aqueous ethanol for 6 hours (70% yield). CA 80,3381(1974) gives the preparation of DMT analogs.

JPS 62,490(1973) gives an indoline synthesis which can be carried one step further (dehydrogenation described in this chapter) to give DMT derivatives. See JOC 41,1118(1976) for indoles from betaketosulfoxides.

Indoles JOC 2,235(1937), TET 27,775,1167(1971); CA 75, 35735,63605,88424,98445,118230(1971), 76,14237(1972); ACS 25,1277(1971); JOC 37,43(1972); CA 77,75123(1972); CPB 20,1395(1972); J. Heterocyclic Chem. 8,903(1972); Diss. Abst. Int. 31B,5871(1971); AP 305,159(1972); JCS (Chem. Comm.) 415(1972).

4-Fluorotryptamines Israel J. Chem. 2,25(1964)

Tryptophan 4-acetic acid from tryptophan JACS 88,3941(1966)

Indoles from o-azidobenzaldehydes JOC 37,719(1972)

Indoles from isatins or oxindoles via diborane Synthesis 2,84 (1972)

Indoles from 2-oxindoles TL 1081(1972)

Indole Derivtives JOC 37,2010(1972)

Psilocin Derivatives Diss. Abst. Int. 32B,2606(1971)

Benzothiophene and Benzofuran TET 28,5397(1972)

Isoindoles TL 4295(1972); JCS(C) 1149(1972)

5-OH-indoles Diss. Abst. Int. 33B,107(1972)

Alkoxyindoles from Hydroxyindoles CA 77,151,914(1972)

MESCALINE AND FRIENDS

Mescaline and related alkaloids are found in varying amounts in cacti of the genera *Lophophora, Gymnacalycium, Stensonia, Mammillaria, Ariocarpus, Opuntia, Trichocereus, Pelecyphora,* and probably others. Members of the Native American Church do quite well with the dried cactus, but extraction of mescaline is desirable since the pure compound seems to produce fewer unpleasant side effects (e.g., nausea). For an excellent review on the occurrence and chemistry of the mescaline type compounds, see JPS 59,1699(1970) (cf. JPS 60,655(1971)). Various species of these cacti occur in southwestern U.S. as well as Central and South America and have been used by the Aztecs and others for millennia. For a good review of peyote see Lloydia 36,1-58(1973).

Extraction of Mescaline from Cactus
JACS 88,4218(1966), Lloydia 29,318(1966)

Dry and grind the top of the cactus and add 50 g to 250 ml methanol and let soak one day. Filter and add 100 ml 1N HCl to the wet powder and let stand about two hours. Meanwhile, evaporate in vacuum the methanol. Filter the HCl solution until clear and add to the residue remaining from the methanol. Make pH about 7 with 2N KOH and add 100 ml $CHCl_3$. Shake in separatory funnel and separate the bottom $CHCl_3$ layer. Add 40 ml water to the $CHCl_3$ and shake in separatory funnel. Dry and evaporate in vacuum the $CHCl_3$ layer to get the mescaline (and other compounds).

The most detailed directions for cactus extraction are contained in PSYCHEDELIC GUIDE TO THE PREPARATION OF THE EUCHARIST by R. Brown, obtainable for $5.50 from L.S.I. Co., PO Box 4374, Austin, TX 78765.

Another source pressure cooked the peyote in acidic water 15 minutes at 15 lbs. and poured off the liquid. This process was repeated 5 more times and the combined water extracts evaporated slowly to a tar which is cooled until able to be formed into small pills. The pills are dipped in Salol (phenyl salicylate), which has been liquified by heating, and allowed to dry. This forms an enteric coating which is said to allow ingestion without nausea. Others merely put the tar in gelatin capsules. Either way the nausea of whole peyote is greatly diminished. Three or four tabs are supposed to have the impact of 15-20 buttons.

Methoxyphenethylamines and Derivatives

Of the nineteen possible methoxyphenethylamines, only three are known to be psychedelic in man at reasonable doses. These are the 3,4,5-trimethoxy-(mescaline), 2,3,4,5 and 2,3,4,5,6 compounds. There are, however, a vast number of possible psychedelic derivatives with methyl, ethyl, Br, and other groups replacing methoxy groups. In addition, innumerable amphetamine derivatives which closely resemble mescaline in their structure and psychedelic properties have been synthesized. This similarity can be seen in the structures of mescaline (R=H) and 3,4,5-trimethoxyamphetamine (R=CH_3).

These amphetamine derivatives have the advantage that they are almost all more active than mescaline, STP being about 80X more active (i.e., requiring only about 5 mg as contrasted with about 400 mg for mescaline), and that they are often easier to synthesize. They do, however, seem to produce somewhat more of the effects of speed (e.g., anxiety, restlessness, sweating) than mescaline. The methyl-enedioxy compounds, on the other hand, produce some of the mildest trips of any psychedelic.

2-5,dimethoxy-4-Br-amphetamine (prepared by bromination of 2,5-dimethoxy-amphetamine -- see JCS 200(1953)) is about ten times more active than STP but it produces a trip similar to that of MDA and is not truly hallucinogenic. The phenethylamine analogs of this compound and of STP are ten times less potent than the corresponding amphetamines and produce only MDA type effects without the intense sensory alterations. Beta-hydroxy-3,4-methyl-enedioxyphenylethyl amine is said to be a very fine trip.

Mescaline (R=H) or 3,4,5-trimethoxyamphetamine (R=CH₃)

When the amino group of amphetamine is methylated, one gets methamphetamine which is considerably more stimulating. The corresponding hallucinogenic analogs of methamphetamine have not yet been shown to have any special stimulating properties, having on the contrary less potency than the unmethylated forms (see JMC

13,134(1970)), but human tests should be done (preliminary results show these compounds to be very pleasant).

N-methyl-MDA is extremely pleasant, producing more euphoria than cocaine, lasting much longer and having a very easy down. N,N-dimethylation or N-ethylation diminish activity drastically and the latter appears to have sedative effects. The methylenedioxy compounds are generally free of marked visual changes, allow considerable voluntary control of the experience and are more potent than the dimethoxy counterparts.

It is probably best to avoid p-methoxyamphetamine (PMA) and 2,5-dimethoxy-4-methylamphetamine (STP), the former because it seems to have a high toxicity and the latter because it lasts too long (e.g., 24 hours for a minimum dose). Other 4-alkyl amphetamines also seem to be toxic. A number of apparent fatalities due to MDA have been noted, but the reports usually involve very large amounts, often in combination with other drugs (e.g., 7 g MDA plus barbiturates) and screening for other, more toxic drugs (in particular, PMA) has not been done.

These amphetamine analogs can be produced by using the appropriately substituted benzaldehyde in place of 3,4,5-tri-methoxy-benzaldehyde, and nitroethane in place of nitro-methane in the aldehyde method for mescaline synthesis (using nitropropane, etc., to give a longer chain results in less activity). However, easier synthetic routes from the naturally occurring (therefore cheap and commercially available) ring substituted propenylbenzenes are given here.

The following table lists common names, position of ring substitution, and approximate activity of the amphetamine derivatives for some readily available allyl and propenyl benzenes (see J. Chromatography 30,54(1967) for further information on these compounds). Activity is relative to mescaline which equals 1 (an activity of 12 means a dose of about 25 mg). Parentheses indicate a methylenedioxy bridge; other substituents are methoxy groups.

Allyl	Propenyl	Substituents	Activity
Safrole	Isosafrole	(3,4)	2
Croweacin	——	2(3,4)	2
Myristicin	Isomyristicin	3(4,5)	2
Dillapiole	Isodillapiole	2,3(4,5)	6
Estragole	Anethole	4	6
——	Asarone	2,4,5	12
——	——	2(4,5)	12
Apiole	Isoapiole	2(3,4)5	12

Amphetamines from Allylbenzenes JACS 91,5648(1969)

To a cooled and stirred solution of 100 ml acetonitrile and 64.8 g mercuric nitrate, add slowly 0.2M of the allylbenzene (keep temperature below 30°). Stir one hour at room temperature, cool and add 200 ml 3N NaOH, then 200 ml 0.5M NabH$_4$ in 3N NaOH. After one hour saturate the water layer with NaCl and extract with ether. Dry and evaporate in vacuum the extract to get the N-acetyl-amphetamines. This procedure may not work with the propenylbenzenes.

If it is desired to remove the N-acetyl group see CJC 51,1407 (1973).

Amphetamines from Propenylbenzenes CA 52,11965(1958)

Use of N-methyl-formamide in place of formamide will give the corresponding N-methyl-amphetamine which is nicer.

Exemplified for 3,4-methylenedioxyamphetamine (MDA).

To a cooled mixture of 34 g 30% H$_2$O$_2$ and 150 g formic acid, add dropwise a solution of 32.4 g (0.2M) isosafrole in 120 ml acetone (keep temperature below 40°). Let stand about twelve hours and evaporate in vacuum. Add 60 ml methanol and 360 g 15% sulfuric acid to the residue and heat on water bath three hours. Cool, extract with ether or benzene and evaporate in vacuum the extract to give 20 g 3,4-methlenedioxybenzyl-methyl ketone (I) (can distill 115/2). Add 23 g (I) to 65 g formamide and heat at 190° for five hours. Cool, add 100 ml H$_2$O$_2$, extract with benzene and evaporate in vacuum the extract. Add 8 ml methanol and 57 ml 15% HCl to residue, heat on water bath two hours and evaporate in vacuum (or basify with KOH and extract the oil with benzene and dry, evaporate in vacuum) to get about 11 g MDA. In this, as in the other syntheses, either the cis or trans (alpha or beta) propenylbenzenes (or a mixture) may be used.

Amphetamines from Propenylbenzenes JMC 9,445(1966)

If the allyl isomer is at hand, it must first be converted to the propenyl as follows (CJC 43,3437(1965)): Add equal weights of the allyl compound and KOH flakes, and absolute ethanol and heat on steam bath or reflux for twenty-four hours; dry and evaporate in vacuum or add two times the volume of water and extract with ether or methylene chloride and dry, evaporate in vacuum (recrystallize-hexane).

0.034M propenylbenzene in a mixture of 3.3 g pyridine and 41 g dry acetone is cooled to 0° and 6.9 g tetranitro-methane added over one minute with vigorous stirring, and stirring continued for two minutes. Add 2.2 g KOH in 40 ml water, add more water and extract the nitropropene with methylene chloride and dry, evaporate in vacuum (recrystallize-methanol). The nitropropenes (which seem to have little activity themselves) can be reduced to the active amphetamines with lithium aluminum hydride or Zn-Hg as described later, or reduced by another method (hydroboration, hydrogenation, Na-ethanol, electrolytic, etc.).

Mescaline and Amphetamines from Styrenes and Propenylbenzenes JACS 86,3565(1964)

The yield of mescaline should be about 50%; that for amphetamines will vary.

0.1M of ring substituted styrene or propenylbenzene in 30 ml tetrahydrofuran in ½L flask. Flush with N_2 and add 33 ml 1 M borane in tetrahydrofuran (see procedure below for preparation). Stir one hour, add 3 ml water and 50 ml 3N NaOH, and then 215 ml 0.31 M fresh chloramine solution (prepared by treating dilute aqueous NH_4OH with Na hypochlorite at 0°; see BER 40,4586 (1907)). Keep at room temperature one hour, acidify with HCl, extract with ether, basify with NaOH and extract with ether and dry, evaporate in vacuum (or just basify and extract with ether and dry, evaporate in vacuum) to get the amine.

To prepare the diborane in tetrahydrofuran, add 0.3 M $NaBH_4$ (or $LiBH_4$) and 0.4 M BF_3 in total 200 ml tetrahydrofuran and keep dry in refrigerator, or generate the diborane in the reaction flask as follows: To a well-stirred suspension of 3.4 g $NaBH_4$ in 150 ml tetrahydrofuran and 0.3M of the styrene or propenylbenzene, add over one hour at room temperature, 15.1 ml BF_3 in ether in 20 ml tetrahydrofuran (keep temperature at room temperature); let stand one hour at room temperature and decompose the excess hydride with water; then add the NaOH and chloramine (or hydroxyl-amino-O-sulfonic acid) and proceed as above to get the amine.

Other references on organoboranes: JACS 82,4710(1960), 88,5853(1966), Org. Reactions 13,28(1963).

Amines from Alkenylbenzenes by Aminoboration BSC 2668 (1973)

Styrenes will give phenethylamines, 1-propenylbenzenes will give amphetamines. The tetrahydrofuran should be dried over KOH pellets and, if desired, distilled from sodium then from lithium aluminum hydride. The diglyme can be vacuum distilled from calcium hydride and stored with calcium hydride. Hydroxylamine-O-sulfonic acid can be purchased or prepared (LAC 702,131(1967); Inorg. Synth. 5,122(1957)). To a 3-necked flask flushed with a nitrogen stream add 0.096M styrene or 1-propenylbenzene in 200 ml diglyme, and then a solution of 1.52 g (0.04M) $NaBH_4$ in 70 ml diglyme. Keep the temperature at 25° and add with stirring over ½ hour 7.3 g (0.052M) 48% BF_3 etherate. Let temperature rise over 3 hours and then reflux 3 hours. Cool and carefully add 11.86 g (0.105M) hydroxylamine-O—sulfonic acid dissolved in 50 ml diglyme. Reflux 3 hours, cool and take up in 10% HCl. Extract with ether, basify the cold acid phase with excess NaOH, extract with chloroform, and dry, evaporate in vacuum to get about 40% yield of the amine.

Amphetamines from Phenylacetates CA 35,5868(1941)

Add 0.44 moles ring substituted phenylacetate, 100 g acetic anhydride and 30 g sodium acetate and heat at 145-150° for 18 hours to get ca. 0.4 moles of the methyl-phenylacetate (I). Add (I) and formamide (or N-methyl-formamide for the N-methyl cpd.), heat 4-5 hours at 180-195°, cool and extract with $CHCl_3$. Evaporate in vacuum, dissolve residue in 40% sulfuric acid and heat at 90-125° for 5-6 hours. Neutralize and add solid NaOH to precipitate about 50% amphetamine. Treat with 10% sulfuric acid to get the sulfate.

Amphetamines from Phenylacetones CA 61,6953(1964)

Exemplified for 2,4,5-trimethoxyamphetamine (I) preparation.

Mix 25 g 2,4,5-trimethoxyphenylacetone, 9.3 g hydroxyl-amine-HCl, 15.6 g K acetate and 400 ml 70% ethanol and reflux 3½ hours. Evaporate in vacuum and extract the residue with 4X150 ml benzene. Wash combined extracts with 2X75 ml water; dry and evaporate in vacuum the benzene (can purify the oil by dissolving in benzene and precipitate by adding petroleum ether) to get aout 20 g precipitate (test for activity). Dissolve 18.1 g precipitate in 200 ml methanol and hydrogenate. Acidify to get about 15 g (I).

Amphetamines from Propenylbenzenes J. Prakt. Chem. 137,345, 138,271(1933), JACS 54,273(1932)

Illustrated for 2,4,5-trimethoxyamphetamine (I) preparation.

Add a saturated solution of 40 g Na nitrite to 10 g asarone in 90 ml ether. Add dropwise over four hours 75 ml 20% sulfuric acid with stirring. Let stand eight hours; filter and wash precipitate with water, ether, ethanol, and dry. Dissolve 10 g precipitate in 60 ml 8% K carbonate in ethanol with stirring and gentle (below 30°) heating. Add 150 g ice, acidify with 100 ml dilute HCl and let stand one-half hour at 0°. Filter, wash with water and dry to get about 7 g yellow crystals (recrystallize-methanol) (2-nitropropenylbenzene). Add 4 g crystals to 100 ml ethanol and 50 ml glacial acetic acid; then add 10 ml concentrated HCl or 50 ml concentrated sulfuric acid for the catholyte in a 40 cm^2 Hg-cathode in a porous cell surrounded by 3N sulfuric acid anolyte with a water cooled lead anode and reduce at 4 amps (about 0.07 amps/cm^2 cathode surface) at 30-40° for twenty hours or until solution is colorless. Evaporate in vacuum to about 20 ml; cool with ice, basify with NaOH and extract with ether. Wash, dry and dilute the ether (or dry, evaporate in vacuum) to get (I).

Note that this provides an alternative to tetranitromethane for nitration of propenylbenzenes, and an alternative to lithium aluminum hydride or Zn-Hg for reduction of nitropropenes.

Amphetamines from Bromobenzenes JACS 63,602(1941)

0.2 M p-methoxy (or other group)-Br-benzene (see this paper for prepartion) and 4.6 g Mg. Rapidly add 18.5 g chloroacetone in 50 ml ether. Evaporate the ether by heating in oil bath and then at about 135° for one hour. Cool and add ice and dilute HCl; extract the oil with ether and dry, evaporate in vacuum to get about 11 g product (can distill 106/18). 0.057 M of the ketone product in 16 g formamide in 100 ml round bottom flask, with an air condenser. Heat twelve hours at boiling and then reflux with 35 ml 30% NaOH eleven hours and separate the amine layer and dry, evaporate in vacuum to get the amphetamine.

The following papers often employ a glass apparatus called a Soxhlet extractor to get maximum yields in a reduction with lithium aluminum hydride; it is, however, unnecessary.

Phenyisopropylamines and Phenethylamines JMC 13,134(1970)

Mix 5.5 g 3,4,5-trimethoxybenzaldehyde (or analog), 2.5 g NH$_4$ Acetate, 25 ml nitroethane (or equimolar amount nitromethane for the analog), 25 ml benzene and reflux about twenty hours; water being removed with a Dean-Stark tube. Cool and wash with 2X25 ml water, 2X25 ml saturated NaHSO$_3$ and 2X25 ml water and dry and evaporate in vacuum the benzene to get (I). Reduce the nitropropene (or nitrostyrene) by any method such as follows: To a stirred suspension of 3 g lithium aluminum hydride in 50 ml tetrahydrofuran, add 4.4 g (I) in 50 ml tetrahydrofuran and reflux one hour. Cool in ice bath and treat slowly with wet tetrahydrofuran until bubbling stops. Filter and evaporate in vacuum (or filter and extract precipitate with 3X25 ml hot tetrahydrofuran and evaporate in vacuum combined tetrahydrofuran) to get the amine.

For a catalytic reduction of (I) see AP 270,340,410(1932) or proceed as follows (AP 273,481(1935)): Dissolve 0.02 M (I) in 250 ml glacial acetic acid and add 10 ml concentrated sulfuric acid and 1 g palladium-carbon (or other catalyst). Reduce at 2 atmospheres H_2 and 15° (about fifteen minutes); filter, add 36 ml 5N NaOH and evaporate in vacuum to get the amine. For Zn-Hg reduction of (I) see below. See also JOC 37,1861(1972) for another catalytic reduction of (I).

Substituted Amphetamines from Substituted Benzenes JACS 68,1009(1946)

4 M of the substituted benzene and 32.4 g dry $FeCl_3$. Cool to -21° in an ice-salt bath and add dropwise with stirring over two hours, 76.5 g allyl-Cl and continue stirring three hours. Add about 2 lbs. crushed ice and 100 ml concentrated HCl. Agitate and separate the organic layer and wash with dilute HCl, then water and filter, dry and evaporate in vacuum (or distill 100-115/10 with Claisen flask, etc.) to get the substituted-1-phenyl-2-Cl-propane (I). Dissolve 0.16 M (about 27 g) (I) in 450 ml ethanol saturated with NH_3 (125 g/L, seal in an iron pipe in autoclave and agitate and heat about nine hours at 160°. Cool and filter and evaporate in vacuum to get the amphetamine in about 20% yield.

Amphetamines from Benzylmethyl ketones JACS 70,1315 (1948)

10 ml water and 0.2 g Pt oxide (or other catalyst) in 300 ml Parr hydrogenation bottle. Shake in H_2 atmosphere ten minutes and add 0.3 M p-methoxy-benzyl-methyl ketone (1-p-methoxy-phenyl-2-propanone) (or analog), 20 g NH_4Cl, 225 ml methanol saturated with NH_3, 25 ml NH_4OH and shake with 1-3 atmospheres H_2 until uptake ceases. Filter, wash precipitate with methanol or acidify with HCl and dry, evaporate in vacuum to get p-methoxy-amphetamine or analog.

For the preparation of phenylacetones from **nitropropenes** see JOC 15,8(1950).

Mescaline J. Chem. U.A.R. 11,401(1968) and many others.

Obtain 3,4,5-trimethoxybenzoic acid or synthesize as follows (Org. Synth. Coll. Vol 1,537(1946)): To a cold solution of 80 g NaOH in 500 ml water in 1 L flask, add 50 g gallic acid; tightly stopper to exclude O and shake occasionally until all acid dissolves. Add 89g (67 ml) dimethyl sulfate, shake twenty minutes, releasing pressure occasionally, and cool to keep temperature below 30°. Again add 89 g dimethyl sulfate and shake ten minutes. Reflux two hours, add 20 g NaOH in 30 ml water and reflux two hours more. Cool, acidify with HCl; filter and wash with water to get 50 g 3,4,5-trimethoxybenzoic acid (recrystallize-2 L hot water) (can recover more by concentrating the filtrate).

A. Do a Fisher esterification by refluxing 100 g 3,4,5-trimethoxybenzoic acid in ethanol with concentrated sulfuric acid for several hours. Cool, filter, to get the ester (I) (recrystallize-ethanol).

A. (Alternative) 100 g 3,4,5-trimethoxybenzoic acid, 20 g NaOH, 55 g NaHCO$_3$ and 300 ml water and add with stirring 94 ml methyl or ethyl sulfate over twenty minutes and reflux one-half hour. Cool, filter, dissolve the precipitate in a small amount hot methanol or ethanol and cool to precipitate (I) (acidify the filtrate to recover unreacted trimethoxybenzoic acid).

B. 1 M (I), 10 M NaBH$_4$; dissolve in methanol, reflux four hours and filter, dry and evaporate in vacuum to get 3,4,5-trimethoxybenzyl alcohol (II).

B. (Alternative) To 4.6 g lithium aluminum hydride in 200 ml ether, add over one-half hour a solution of 23 g (I) in 300 ml ether. Slowly add 50 ml ice water; decant the ether and add 250 ml ice cold 10% sulfuric acid. Extract with 3X50 ml ether and dry, evaporate in vacuum (can distill 135-137/0.25) to get (II).

C. To a stirred solution of 39.6 g (II) in 150 ml CHCl$_3$ (or CCl$_4$ or benzene), at 0-5°, add 35 g SOCl$_2$ in CHCl$_3$ over one-half hour; let warm to room temperature and stir for one hour. Evaporate in vacuum and pour residue on ice cold water. Filter and wash precipitate 3X with water to get the trimethoxybenzyl-Cl (III) (recrystallize-benzene-hexane).

C. (Alternative) 25 g (II) and 125 ml ice cold concentrated HCl is shaken vigorously until a homogeneous solution is obtained. Let stand four hours until gummy precipitate forms and dilute with 100 ml ice water. Decant the water and extract with 3X50 ml benzene. Dissolve precipitate in this benzene, wash benzene with water and dry, evaporate in vacuum. Suspend the residue in a small amount ice cold ether, filter through cold funnel and wash with cold ether. Let filtrate stand in refrigerator to get more precipitate of (III) (recrystallize-benzene).

D. 43.2 g (III), 40 g KCN, 300 ml 85% formic acid (or DMSO, acetonitrile, 70% methanol, but lower yield for these). Add 150 ml water and reflux two hours. Cool, filter, and extract filtrate with 3X50 ml $CHCl_3$. Wash $CHCl_3$ with 3X50 ml water and dry, evaporate in vacuum to get the phenylacetonitrile (IV) (recrystallize-benxene-hexane, yield about 34 g).

E. HCA 53,50(1970) (see JMC 14,375(1971) or the following for other methods of reducing (IV)). 40 g (IV) in 90 ml methanol and 10 g aqueous 50% Raney-Ni (or substitute catalyst described in chemical hints section). Add dropwise over ten minutes with vigorous stirring a solution of 7.6 g $NaBH_4$ in 25 ml 8N NaOH and cool to keep temperature at 50°. H_2 evolution stops in about five minutes. Filter, wash with methanol and evaporate in vacuum to get mescaline (V).

E. (Alternative) Dissolve 10 g lithium aluminum hydride in 300 ml tetrahydrofuran or ether. Stir and add 20.4 g (IV) in 60 ml tetrahydrofuran or ether over one-half hour. Reflux three hours and cool. Slowly (dropwise at first) add 10 ml concentrated sulfuric acid in 40 ml water; separate the aqueous layer and basify with concentrated NaOH. Filter off any precipitate and extract brown oil with 3X30 ml ether. Wash ether with water, dry and add 1 ml concentrated sulfuric acid and 25 ml ether. Filter and wash the white precipitate with ether to get (V) (recrystallize-ethanol). (Alternatively, add 1.5N sulfuric acid dropwise until bubbling stops, filter (wash precipitate with solvent and add washings to filtrate), and dry, evaporate in vacuum (or bubble HCl gas through, or add a little concentrated HCl and evaporate in vacuum) to get (V)).

Mescaline and Amphetamines via Aldehydes J. Chem. U.A.R. 11,401(1968) and many others.

Illustrated for mescaline.

A. To 39.6 g trimethoxybenzyl alcohol (see preceding method for preparation) (or other analog) in 250 ml methanol, stirred and cooled to 0°, carefully add 54 ml Br_2 over one hour at 0°. Let temperature rise to room temperature, stir for two hours and add 30 ml saturated Na thiosulfate. Filter off the aldehyde (I) (recrystallize-benzene).

A. (Alternative) To 11 g hexamethlenetetramine in 70 ml $CHCl_3$, add 18 g trimethoxybenzyl-Cl (or other analog) and reflux four hours. Evaporate in vacuum and add 35 ml acetone to cause precipitation. Filter and heat precipitate with 100 ml water for twenty minutes. Add 17.5 ml concentrated HCl and reflux five minutes. Evaporate in vacuum the acetone, cool and filter to get (I).

B. To 40.4 g (I) and 16 g nitromethane (or equimolar amount nitroethane for amphetamines) in 100 ml methanol, add 14 ml 3% methylamine in methanol (under N_2 if possible) and let stand at room temperature one day. Cool to -10°, filter, wash precipitate with cold methanol and dry in vacuum or warm oven to get the nitrostyrene (or nitropropene) (II), which can be reduced as described previously, or as in step C.

B. (Alternative) 40.4 g (I) and 16 g nitromethane (or equimolar amount nitroethane) in 150 ml glacial acetic acid and 15 g NH_4 acetate. Reflux 1½ hours; cool, filter, and recrystallize from acetic acid or methanol to get (II).

C. CPB 16,217(1968). Suspend 0.2 M (II) and Zn-Hg from 200 g Zn and 20 g $HgCl_2$ in 2 L ethanol and add with vigorous stirring portions of concentrated HCl until the yellow color disappears. Continue stirring one-half hour, filter, evaporate in vacuum to get about 0.14 M of the amine.

Phenethylamines from Acetophenones JOC 22,331(1957)

Acetophenones can be obtained by various routes such as by reacting acetic anhydride with the substituted benzene (Shirley-PREPARATION OF ORGANIC INTERMEDIATES, 1951, pg. 190) and may then be isomerized, if necessary, with aluminum chloride (JCS 232(1944)).

A mixture of 0.33M of the acetophenone, 39 g of redistilled morpholine and 14.4 g sulfur is refluxed for 12 hours and the wrm solution poured into 175 ml hot ethanol. Cool to precipitate about 80% yield of the substituted phenylacetothiomorpholide (I). Mix about 50 g (I), 110 ml acetic acid, 16 ml sulfuric acid and 25 ml water and reflux 5 hours. Decant from the small amount of tar formed with stirring into 850 ml water. Filter, wash the precipitate with water and heat the precipitate with 225 ml 5% aqueous NaOH. Filter and acidify the filtrate with dilute HCl to give about 80% yield of the substituted phenylacetic acid (II). Mix about 21 g (II) and 25 g phosphorus pentachloride (caution), and after the initial reaction subsides, warm on the steam bath 10 minutes. Distill under reduced pressure to remove the POCl and gradually add the residue to ice cold concentrated NH_4OH. Filter, wash precipitate with water and air dry (can recrystallize from benzene with a little ethanol added) to get about 18 g (85%) of the substituted phenyl-acetamide (III). (III) may be reduced by various means such as the following. To a stirred suspension of 8.6 g lithium aluminum hydride in 500 ml dry ether add a solution of 10 g (III) in 600 ml boiling reagent benzene, adding additional hot benzene to redissolve any precipitate. Stir and reflux for 22 hours and then hydrolyze by adding water cautiously and 10% sulfuric acid. Filter, heat the precipitate with concentrated HCl to dissolve and cool to preicipitate the substituted phenethylamine.

Phenylethylamines from Benzaldehydes Indian J. Chem. 5,471 (1967), JMC 11,534(1968)

A. 0.02 M substituted benzaldehyde, 27 g acetic anhydride, 12 g fused K acetate in 250 ml round bottom flask with air condenser and $CaCl_2$ tube. Heat on oil bath at 160° one hour and then at 175° for four hours. Pour into water (neutralize with Na carbonate and steam distill to recover unreacted aldehyde), cool, acidify and filter to get about 60% yield substituted cinnamic acid (I).

B. Dissolve 10 g (I) in 100 ml water containing 10% NaOH and add with stirring, 300 g 3% Na-Hg portionwise at intervals of one hour and stir for ten hours. Filter, concentrate to one-half volume and acidify to precipitate the hydrocinnamic acid (II) in about 50% yield.

C. 10 g (II) and 15 g $SOCl_2$ are refluxed on water bath two hours and evaporated in vacuum on water bath. Pour residue into cold liquid NH_3 or NH_4OH with stirring. Filter, wash precipitate with water to get about 80% yield substituted hydrocinnamamide (III) (recrystallize-benzene or dilute ethanol).

C. (Alternative) To 0.1M ethyl-chloroformate in 100 ml $CHCl_3$ at -30° add a cold solution of 0.1 M (II) and 0.1 M triethylamine in 100 ml $CHCl_3$ over forty minutes. Stir 1½ hours at -20° to 5° and bubble NH_3 through the cold mixture for twenty minutes. Stir one-half hour at room temperature, filter and extract the solid with $CHCl_3$. Combine $CHCl_3$ extracts and filtrate and wash two times with cold NaOH solution and two times with water. Dry and evaporate in vacuum to get (III).

D. Pass the chlorine from 3 g $KMnO_4$ and excess HCl into 120 ml 10% NaOH to make a solution of hypochlorite (Clorox may do). Add 10 g finely powdered (III) and stir to dissolve amide, while warming to 50°. Heat one hour at 85°, add 30 g KOH and heat two hours. Separate the oil and extract the aqueous layer with ether. Add ether to oil and dry, evaporate in vacuum to get the substituted phenylethylamine (IV).

D. (Alternative) To a stirred and cooled solution of 0.1 M NaOH in 100 ml water at -5° add 0.04 M bromine over five minutes and stir one-half hour at 0°. Add 0.02 M (III) and stir 1½ hours at 0-5°. Stir sixteen hours at room temperature and 70° for one hour. Cool and extract with ether, dry and evaporate in vacuum to get (IV).

Phenylethylamines from Substituted Benzenes CA 57,6721(1963)

0.1 M m-methoxy-toluene (or analog), 30 ml HCl, 30 ml benzene; cool to 0° and with cooling and stirring saturate with HCl. Add 0.15 M formaldehyde and then more HCl for fifteen minutes. Stir at room temperature for two hours and dry, evaporate in vacuum the benzene layer to get about 30% 2-methyl-4-methoxy-benzyl-Cl (or analog). 0.5 M (I), 0.8 M NaCn, 5 g NaI or KI in 250 ml dry acetone; stir and boil twenty hours. Filter, wash precipitate with 100 ml acetone and dissolve precipitate with 75 ml benzene. Wash with 100 ml water and dry, evaporate in vacuum to get about 80% yield 2-methyl-4-methoxy-benzyl-CN (II). 0.1 M (II) in 60 ml 10 N NH_3 in methanol; 4 g $Ni-Cr_2O_3$ and shake in autoclave at 70-100 atmospheres H_2 and 120° to get about 80% yield of the phenethylamine.

Phenylethylamines from Phenols CA 68,86944(1968)

To 5.1 g omega-phthalimidoethyl-Br and 3.1 g 2,6-dimethoxy-phenol (or analog) in 500 ml ether add 2.66 g dry $AlCl_3$ portionwise and let stand twenty-four hours at room temperature. Heat two hours at 60°, filter, wash to recover starting material and dry, evaporate in vacuum to get 25% yield of (I). To an ether solution of diazomethane from 8 g $H_2N-CON(NO)CH_3$ add 4 g (I) portionwise and filter, evaporate in vacuum to get about 100% yield of (II). Heat 2 g (II) in 20 ml aqueous HCl for four hours and cool, precipitate, filter, evaporate in vacuum the filtrate to get about 30% yield mescaline.

Phenethylamines JMC 15,214(1972)

This is a rapid, convenient procedure. If trimethoxybenzyl alcohol is used in place of p-anisyl alcohol, mescaline will result. Shake 100 g p-anisyl alcohol (or 0.72 moles analog) with 500 ml concentrated HCl for 2 minutes. Wash the organic phase with water, 5% $NaHCO_3$ and water, and then add over 40 minutes to a stirred slurry of 49 g NaCN in 400 ml dimethylsulfoxide with ice water cooling to keep temperature at 35-40°. After completing addition, remove cooling bath, stir for 90 minutes and then add to 300 ml water. Separate the small upper layer and extract the aqueous-DMSO layer with 2X100 ml ether. Add the ether extracts to the small upper layer, wash once with water and dry ($MgSO_4$). Add 600 ml ether to a dry flask and chill in ice as 80 g of anhydrous $AlCl_3$ is added portionwise, followed by 23 g of lithium aluminum hydride. Add the washed and dried ether extracts at such a rate as to give gentle reflux without external heat (ca. 1 hour). Stir for 2 hours, ice chill and treat dropwise with 25 ml water then 250 ml 20% aqueous NaOH, with periodic addition of ether through the condenser to replenish losses and facilitate stirring. Filter and wash precipitate with ether; add ether to filtrate and mix with one third its volume of absolute ethanol. Slowly and with continuous swirling and ice cooling add 60 ml concentrated HCl. Cool to 0° and filter to get the amine hydrochloride in ca. 75% overall yield (or can obtain the crude amine by filtering and drying, evaporating in vacuum the final ether solution). Note the use of concentrated HCl to obtain t.:e crystalline end product in contrast to many other papers which employ HCl gas.

N-Methylation of Amines JMC 13,134(1970), 15,214(1972)

Either phenethylamine or phenylisopropylamine will work. If the hydrochloride salt is available, this should first be changed to the free amine by stirring with concentrated aqueous NaOH.

Treat 0.54 moles of the amine with 100 ml of benzene and 70 g of benzaldehyde in a flask fitted with a Dean-Stark trap and reflux until no more water is present in the condensate (ca. 1 hour) (alternatively, reflux ½ hour and then distill until the temperature reaches 100°): remove the trap and add a solution of 82 g dimethylsulfate in 200 ml benzene through the condenser at a rate sufficient to maintain reflux (ca. 15 minutes). Reflux 30 minutes (or heat 90 minutes on a steam bath), add 200 ml water and reflux (or heat on steam bath) for ½ hour more. Separate the aqueous layer, extract it twice with ether or benzene, basify with 50% NaOH and extract twice more. Add the extracts to the organic layer and dry, evaporate in vacuum to get ca. 70% yield of the N-methyl-amine. Can crystallize by dissolving in 500 ml 20% absolute ethanol-ether and treating with 50 ml concentrated HCl while swirling and cooling. The crystals can be further purified by washing with ice cold 20% ethanol-ether and recrystallizing from ethanol.

N-methylation of hallucinogenic amphetamines seems to produce a very smooth, mellow, euphoric trip, and the same is probably true for phenylethylamines.

For N-methylation via the imides see JOC 38,1348(1973). N-methylation appears to decrease the potency. Larger N-alkyl groups seem to have a sedative action.

N,N-Dimethylamines JMC 13,134(1970)

To 14 g formic acid, cooled in an ice water bath, add dropwise 0.016 mole of the amine and then 3.6 g of formaldehyde (ca. 10 ml of formalin) and reflux for 5 hours. Cool to room temperature, add 7 ml concentrated HCl and evaporate in vacuum. The resulting oily dimethylamine can be purified by dissolving in 25 ml water, extracting with 2X25 ml $CHCl_3$. Basify the aqueous layer with 2N NaOH and extract with 3X25 ml ether, and proceed as described above for the N-methylamines. This procedure should work for both phenethylamines and phenylisopropylamines, and should affect the trip similarly to N-monomethylation.

See the following method for another easy route to N-alkylation which will probably work for dialkylation.

Although N,N dialkyl compounds have very little activity, they are probably worth further investigation.

Amphetamines via phenylacetones CT 3,313(1968) CF. JACS 58,1808(1936) and JCS 18(1930)

Note that the following procedure leads from the aldehyde to the amine or N-substituted amine without involving a troublesome catalytic hydrogenation for reducing the intermediate nitropropene.

Mix 0.25M substituted benzaldehyde, 0.3M nitroethane, 50 ml dry toluene and 5 ml n-butylamine (or other amine), and reflux 3 hours with a Dean-Stark trap (or prepare the nitro-propene as described elsewhere here). Add 50 g iron powder and 1 g $FeCl_3$ (optional) and reflux while adding 90 ml concentrated HCl over 3 hours. Reflux 1 hour more, add 2 liters of water and extract 3 times with ether, then dry and evaporate in vacuum (or steam distill until about 3.5 liters of distillate is obtained; extract the distillate 3 times with toluene; wash the toluene layers with 7 g $NaHSO_3$ in 225 ml of water, then 3 times with water and dry, evaporate in vacuum) to get the ketone. Mix 0.13M ketone, 28 g formamide (or dimethyl-formamide if the N,N-dimethylamine is desired) and 3 ml formic acid and heat at 160°. Add 3 more ml formic acid and heat 16 hours at 170-180° adding formic acid from time to time to keep the pH acid. Distill off the water formed (about 16 ml), cool and extract with 3X70 ml benzene. Distill off the benzene and reflux the residue 7 hours with 30 ml concentrated HCl. Chill, basify with 10% NaOH and extract with 3X70 ml ether. Dry and evaporate the ether in vacuum to get the amine. Since some amines are unstable to heating, it is perhaps best just to dry and evaporate the benzene extract.

To obtain an N-substituted amine reflux 0.1M ketone, 17 g aluminum filings or foil, 50 ml ethanol, 40 ml 30% aqueous n-butylamine (or other amine), and 0.5 g of mercuric chloride for 3 hours. Cool and pour on 500 g crushed ice and 200 ml 10% KOH. Extract 3 times with ether and dry, evaporate in vacuum (or wash combined ether layers 2 times with 10% HCl, basify acid extract with 15% NaOH and extract 3 times with ether and dry, evaporate in vacuum) to get the n-butyl-amine (or other amine).

Alkoxybenzaldehydes

2,4,6-Trimethoxybenzaldehyde JCS 4964(1952)

This method should also work for 3,4,5-trimethoxybenzene and other symmetrically substituted benzenes.

To 42 g 1,3,5-trimethoxybenzene, 30 g formanilide and 200 ml ether add 19 g $POCl_3$ and let stand twelve hours. Evaporate in vacuum, add 900 ml 5% NaOH and steam distill residue to get the aldehyde (recrystallized benzene-ligroin).

For a different synthesis see Org. Synth. 2,11(1940).

Methylenedioxybenzaldehydes

Substituting methyl sulfate for methylene sulfate will probably give the dimethoxybenzaldehydes.

Dissolve 4 g 3,4-dihydroxybenzaldehyde (or any other benzaldehyde having adjacent OH groups) and 7.6 g KOH in 50 ml water. Heat to 50° (under N_2 if possible) with stirring and add 5 g methylene sulfate; stir and heat two hours. Add 500 ml water and sufficient K sulfate to dissolve the precipitate. Extract with ether, acidify the aqueous phase with HCl to precipitate unreacted aldehyde, and dry, evaporate in vacuum the ether to get the title compounds (recrystallize-methanol).

2,4,5-Trimethoxybenzaldehyde JACS 57,1126,2739(1935)

To 13 g 3,4-dimethoxyphenol in 20 ml methanol add a solution of 5.3 g KOH in 100 ml methanol and then 12 g methyl iodide. Reflux two hours, add 300 ml water and make pH 10 with 5% NaOH. Extract with ether and dry, evaporate in vacuum the pooled extracts to get a clear oil. Mix 17.3 g N-methyl-formanilide and 19.6 g $POCl_3$ and let stand at room temperature for one-half hour. Add 8.5 g of the oil, heat two hours on steam bath (or boiling water) and pour black viscous product on 800 g cracked ice and let stand three hours, or until no more precipitation to get title compound. Filter (recrystallize-methanol).

Simple methods of preparing phenylacetonitrile from diazoacetonitrile and triphenylboron are given in JACS 90,6891(1968) and from the borophenyl-9-borabicyclo (3.3.1) nonane (BBN) and Cl-acetonitrile in JACS 91,6854(1969). These methods should work with substituted phenyl compounds to give mescaline and analogs. Likewise (JACS 90,5936(1968)), a preparation of phenylketones from triphenylboron or BBN and diazoacetone, or bromo-acetone (JACS 91,6852(1969)) may be used to prepare the phenylacetones which can be reduced to the amphetamines in one step as already described. For synthesis of BBN see JACS 91,4304(1969) and for borabicyclononane see JACS 90,5280(1968).

The amphetamine and mescaline analogs which can be obtained by replacing carbon atoms with O, N, or S atoms may also be active, but little work has been done on this. For the synthesis of thienyl analogs see JACS 64,477(1942).

Other Syntheses

N-Methyl Mescaline HCA 35,1577(1952)

p-Methylephedrine Pharmazie 24,735(1969)

4-Nitropropenylbenzene GCl 100,846(1970)

Amphetamines and Phenylethylamines JMC 11,186(1968), 13, 26,134(1970); JOC 19,11(1954), 20,102,1292(1955), 22,331 (1957), 23,1979,2034(1958), 25,2066(1960); Alab. J. Med. Sci. 1,417(1964); JCS 18(1930); BSC 835(1962); BER 63,3029 (1930), 65,424(1932), 67,696(1934); CA 6953(1964); Arznei. Forsch. 9,157(1959); JACS 51,2262(1929), 63,602(1941), 76,5555(1954); Pharmazie 22,19(1967); CT 3,313(1968).

Bromoamphetamines JMC 15,413(1972)

N-substituted mescaline derivatives and methamphetamine analogs Diss. Abst. Int. 32B,5704,5706(1972).

Methamphetamine analogs JMC 16,14(1973), MMDA CJC 46,75 (1968).

See CT 8,308(1973) for a route from phenethyl or phenyl-propylchlorides to the amines. For older MDA syntheses see JACS 62,425(1940); JCS 15(1943), 1527(1951).

For other compounds related to STP see CJC 51,1402(1973), and

Psychopharmacology Communications 1,93(1973)
MDA Japanese Patent 8573 (5 Oct 1956)

HARMALINE AND OTHER BETA-CARBOLINES

Although some of the compounds of this group occur in plants used as hallucinogens for millenia by the aborigines of South America and elsewhere, very little scientific work has been done on them. The *Homa* of Asia Minor appears to be syrian rue (*Peganum harmala*) from which harmaline was first isolated. Harmaline (1-methyl-7-methoxy-3,4-dihydro-beta-carboline) is about three times less potent than the 6-methoxy isomer, and very probably the 5-methoxy isomer will be even more active. The 6-methoxy isomer is also called 10-methoxy-harmalan. Activity appears to be greater when the methoxy group is replaced by a longer chain alkoxy group. If a double bond is also placed in the 3-4 position of harmaline, one obtains harmine. These compounds are active at about 200 mg orally. Most of these compounds are probably legal.

Harmaline

Harmaline and Isomers CJC 37,1856(1959)

Dissolve 3.3 g 4,5, or 6 methoxy (or ethoxy, methyl, etc.)-tryptamine or its HCl salt in 350 ml 0.1 N HCl: heat on steam bath two hours with 1.1 g glycoaldehyde (reaction over when aliquot no longer gives a precipitate with dinitrophenylhydrazine). Filter, concentrate by heating on water bath or evaporate in vacuum; basify with 20% NaOH and extract with ether (best to do in an extractor for eighteen hours). Dry and evaporate in vacuum the extract to get about 5 g residue or oil which may precipitate on standing. Add 250 ml 90% phosphoric acid and heat on steam bath two hours. Evaporate in vacuum (or dilute with water, basify with 20% NaOH; extract with ether in extractor and dry, evaporate in vacuum the extract) to get about 2.5 g harmaline (or analog).

Norharmaline and Isomers JACS 72,2962(1950)

To 24 ml acetic anhydride in 80 ml formic acid add 22 g 4,5 or 6 methoxy (or analog) tryptophan and reflux ½ hour. Concentrate to thick syrup on steam bath with vacuum and add slowly with stirring 180 ml water and let stand twelve hours at 0°. Filter, wash with 2% HCl and water to get about 18 g N-formyl-methoxy-tryptophan (I) (recrystallize-benzene-ethanol). 10 g (I), 119 g polyphosphoric acid, 10 ml POCl₃ mixed in 500 ml flask with vigorous stirring. Heat on oil bath at 125° for eighty minutes (HCl and CO_2 evolution). Add ice, cool, filter and neutralize filtrate with concentrated NH_4OH to get about 5 g norharmaline or isomer. To get harmaline isomers, use 1.4 g N-acetyl-methoxy-tryptophan, 14 g polyphosphoric acid and 3 ml POCl₃ and proceed as above.

Harmaline and Isomers JACS 70,223(1948)

Dissolve 27 g 4,5, or 6 methoxy (or ethoxy, methyl, etc.) tryptophan in 50 ml freshly distilled acetaldehyde and 1 L water and heat at 50° in loosely stoppered flask three hours. Heat on steam bath five hours to remove acetaldehyde, then add 5 L water and heat to boiling. Add 1.2 L 10% K dichromate and 240 ml glacial acetic acid and continue heating three minutes. Cool and add excess Na sulfite; take pH to 8 with Na carbonate Extract with 5 L ether and dry, evaporate in vacuum (or simply evaporate in vacuum after cooling) to get the harmaline isomer.

1,2-Dihydroharmaline JPS 57,269(1968)

Dissolve 4 g glyoxylic acid monohydrate in 220 ml water and add with stirring to a solution of 10 g 4,5, or 6 methoxy-tryptamine-HCl in 80 ml water. Adjust pH to 4 with 10% KOH and stir five hours at room temperature. Filter and wash with water to get about 4.5 g precipitate. Melt to decarboxylate or to 4 g precipitate add 30 ml concentrated HCl and 100 ml water and heat at 65° one hour. Add water to dissolve the solid and add excess 10% KOH. Filter to get the title compound or analog.

Harmaline analogs BSC 2262(1962)

Brominate cyclohexanone-2-COOH (JACS 72,2127(1950)) and cover to the ethyl ester. Add 1M ethyl-3-Br-cyclohexanone-2-carboxylate to 2M p-methoxy-aniline in cold benzene. Recrystallize the product from ether-petroleum ether and reflux sixteen hours in presence of ZnCl₂ in dry ethanol. Filter, evaporate in vacuum and recrystallize-petroleum ether. Dehydrogenate with 5% palladium-carbon and then treat with NaOH (JCS 530(1945)). Heat at 180° to get the analog of 6-methoxy-norharmine.

Sulfur Analogs of Harman and Other beta-Carbolines JACS 72,4999(1950)

To synthesize 6-methoxy-3,4-dihydro-beta-carboline (10-methoxy-harmalan), add 5-methoxy-tryptamine to acetic anhydride and let stand twelve hours at 10°. Dilute with water, basify and extract with methylene Cl and dry, evaporate in vacuum to get melatonin (I). Reflux (I) in xylene in the presence of P_2O_5 to get the title compounds. *Other References:* JMC 7,136(1964); JPS 57,1364 (1968), 59,1446(1970); JACS 70,219(1948); JCS 1602(1921), 1203(1951), 4589,4593(1956); Organic Synthesis 51, 136 (1971). For the preparation of tryptophan-4-acetic acid from tryptophan see JACS 88,3941(1966).

Sulfur isosteres of harmine, harmaline, etc. J.Het. Chem. 9,1265 (1972). *Harmaline and analogs* BSC 2058(1973). *Beta-carbolines from sugars and tryptamines in 1 step* BER 106,2943(1973).

MUSCIMOLE AND OTHER ISOXAZOLES

It appears likely that mushrooms were the source of man's first psychedelic experiences. The "soma" praised in the Rig Veda of ancient India (about 2,000 BC) is now thought to be a mushroom (the fly agaric, Amanita muscaria, and possibly other Amanita species). Much of the Eurasian religion and mythology may have had its origin in the psychedelic properties of these mushrooms. Probably all members of the genus Amanita are psychedelic, but ingesting them is unwise since there is probably not a great difference between the psychedelic and lethal dose of mushroom for some species. Amanita muscaria itself is seldom lethal but be very careful if you eat it. It is said to be active if dried, powdered, and smoked. In the USA it fruits in early summer and again in late summer (in fall and winter on the Pacific coast). Since muscimole is very water soluble, it can be extracted from the dried, powdered mushroom (or perhaps even fresh ones) by soaking in water (hot or even boiling water may be OK) and straining -- but drink cautiously. Most of the muscimole is contained in the red "skin" and the underlying yellow tissue. Some Russian peoples regarded thorough drying of Amanita as a necessary preliminary to ingestion. It has recently been suggested that there are significant chemical differences between New and Old World A. muscaria.

The main psychedelic component of these mushrooms appears to be muscimole (5-aminomethyl-3-OH-isoxazole), which is active in an oral dose of about 10 mg. See HCA 48,920(1965) for extraction of muscimole from the fungus.

$$H_2N-CH_2 \quad \overbrace{}^{OH} \quad N \quad O$$

Muscimole

The value of muscimole as a psychedelic is diminished by the dizziness and muscle twitching which seem to occur (at least in some people), but a small dose in conjunction with another psychedelic should be very interesting. Ibotenic acid also occurs in Amanita, and though not itself a desirable psychedelic, it can be converted to muscimole by dissolving in dimethylsulfoxide or refluxing in water.

Its synthesis is generally more difficult than that of muscimole (see CA 65,2266(1966) and CPB 15,1025(1967), 19,46(1971)). Several routes for the synthesis of muscimole (also called agarin or pantherine) follow. For a good review of this subject see Prog. Chem. Org. Nat. Prod. 27,262(1969). Muscimole is probably legal everywhere.

Muscimole Synthesis

Method 1 Rend. Ist. Lomb. 93A,150(1959), 94A,735(1960), JCS 172(1968)

Dissolve 100 g 1-Cl-3-Butanone (JACS 75,5438(1953)) in 300 ml dimethylformamide; stir and add 77.5 g $NaNO_2$ with cooling. Stir five hours and let stand twelve hours. Add 300 ml water; extract with ether and dry, evaporate in vacuum the extract (can distill residue 86/2) to get 50 g 1-NO_2-3-butanone (I). Dissolve 18 g (I) in 90 ml 48% HBr and reflux ten minutes. Add 100 ml water and steam distill. Dry, extract with ether and dry, evaporate in vacuum the extract (can distill 67-8/18) to get 4.8 g 3-Br-5-methyl-isoxazole (II). 32.4 g (II) and 85.5 g KOH in 220 ml methanol and reflux twenty four hours (under N_2 if possible). Add 1 L water, extract with 4X250 ml ether and dry and evaporate the combined ether extracts to get 13 g liquid 3-methoxy-5-methyl-isoxazole (III). Add 100 ml n-butyllithium in hexane (2 moles) with stirring over twenty minutes below -65° to 22.6 g (III) in 100 ml dry tetrahydrofuran. Stir at -75° one hour and pour onto a stirred mixture of 1 kg dry ice (solid CO_2) and 1 L ether. Let stand twelve hours or until CO_2 evaporates and extract with 3X200 ml 2N NaOH. Acidify with dilute HCl and extract with 3X100 ml ether. Dry and evaporate in vacuum the extract to get 25 g 3-methoxy-isoxazole-5-acetic acid (IV) (recrystallize-petroleum ether-ethyl acetate). Dissolve 5.2 g (IV) in 100 ml 3% HCl in methanol and reflux three hours. Evaporate in vacuum and recrystallize-petroleum ether (or dissolve residual oil in ether, wash with $NaHCO_3$ and dry, evaporate in vacuum) to get 4.5 g of the methyl ester (V) (can distill 80/0.1). Add 6.5 g (V) in 20 ml methanol to 3 g hydrazine hydrate in 20 ml methanol and reflux four hours. Evaporate in vacuum and dry in exsiccator over concentrated sulfuric acid for twelve hours to get 6 g 3-methoxy-isoxazole-5-acetohydrazide (VI) (recrystallize-ethyl acetate). Dissolve 5.13 g (VI) in 50 ml dry tetrahydrofuran and treat with 50 ml 1.75 N HCl in tetrahydrofuran. Cool to -40° and add over ten minutes with stirring 5 ml tertiary butyl nitrite in 20 ml tetrahydrofuran (temperature rises to about 10°). Evaporate in vacuum and dissolve the yellow oil in 100 ml ethyl acetate. Wash with 50 ml NaCl containing 1% $NaHCO_3$ and dry, evaporate in vacuum to get the azide. Dissolve the azide in 50 ml toluene and heat at 100° until N_2 evolution stops (about thirty minutes). Add 10 ml ethanol and reflux two hours. Evaporate (room temperature/15) to get 4 g 5-(ethoxy-carbonylamino)-methyl-3-methoxy-isoxazole (VII) (can distill 129/0.2). Test for psychedelic activity. Dissolve 4 g (VII) and 2.5 g KOH in 12 ml ethanol and reflux eight hours. Dissolve in 20 ml water, acidify with dilute HCl and evaporate to dryness. Dissolve residue in hot ethanol and evaporate to get 3 g 5-aminomethyl-3-methoxy-isoxazole (VIII). Test for activity. 1 g (VIII) in 10 ml glacial acetic acid and 4.5 g HBr and reflux one hour. Evaporate in vacuum to get muscimole.

Method 2 JACS 62,1147(1940), TL 2077(1963), CA 65,2267 (1966)

Pass chlorine gas into an ice cold, well-stirred solution of 5 ml acetylketene in 30 ml CCl_4 until there is a 4.5 g increase in weight (solution is slightly yellow). Pour slowly into excess methanol or ethanol at 0° and distill at 118/17 to get 6 ml methyl (or ethyl)-4-Cl-acetoacetate (I). To 2.7 ml methanol saturated with dry HCl at 0°, add a mixture of 10 g (I), 20 g methyl orthoformate (trimethoxy-methane) and 13 g methanol and reflux four hours. Pour hot into 200 ml ice water and adjust pH to 8 with 30% NaOH. Extract four times with ether and evaporate and distill to get methyl (or ethyl)-4-Cl-3,3-dimethoxy-butyrate (II). Dissolve 40 g (II) in 20 ml methanol and add hydroxylamine.HCl in methanol. After ninety-six hours at room temperature (under N_2 if possible), evaporate in vacuum. Can purify the residue by dissolving in water and put on anionic column; wash column to neutrality and elute with 2N acetic acid; just before the acid elutes, the alkaline fraction giving a positive $FeCl_3$ test appears; evaporate in vacuum this fraction below 40° and dry at 40/0.5 for twelve hours. Dissolve 5 g product in 130 ml glacial acetic acid and saturate at room temperature, then at 0° with dry HCl. Let stand sixteen hours at room temperature and evaporate in vacuum at 40°. Dilute with water and evaporate three times. Extract with 2X130 ml hot ether and filter, evaporate in vacuum to get 3-Cl-methyl-5-OH-isoxazole (III) (recrystallize-acetone). Heat (III) sixteen hours at 90° in concentrated NH_4OH in autoclave and evaporate to get muscimole.

Method 3 GCI 91,61(1961), JCS 172(1968)

To 1 g 3-Cl-5-methyl-isoxazole in 14 ml 1.4 N NaOH add with

stirring 1.1 g $KMnO_4$ in 20 ml water. Filter, decolorize with $NaHSO_3$ and acidify with dilute sulfuric acid. Extract with ether and dry, evaporate in vacuum to get 0.8 g 3-Cl-isoxazole-5-COOH (I). To 2.5 g (I) in 7.5 ml water add 15 g KOH in 43 ml methanol and reflux four hours (under N_2 if possible). Cool, acidify with concentrated HCl and extract with ether five times. Dry and evaporate in vacuum to get 2 g oil which precipitates to give 3-methoxy-isoxazole-5-COOH (II). Dissolve 5.2 g (II) in 100 ml 3% HCl in methanol and reflux three hours. Evaporate in vacuum and recrystallize-petroleum ether (or dissolve residual oil in ether, wash with $NaHCO_3$ and dry, evaporate in vacuum) to get the methyl ester (III). Add with stirring 0.8 g (III) to 40 ml aqueous NH_3 (density about 0.88) and stir thirty minutes at room temperature. Filter, wash with cold water and dry to get 0.5 g 3-methoxy-isoxazole-5-carboxamide (IV). Dissolve 37.8 g $NaBH_4$ in 100 ml diglyme (dimethyl ester of diethylene glycol) and 23 ml BF_3 etherate in diglyme; add to 4.6 g (IV) in 100 ml tetrahydrofuran and reflux forty-eight hours. Add HCl, evaporate in vacuum and dissolve residue in water. Basify with 50% KOH and extract with ether. Dry, filter, evaporate in vacuum or treat with HCl to give (VIII) of method 1 which converts to muscimole as described.

Method 4 CA 65,2266(1966)

Dissolve 626 g 2-(2-nitrovinyl) furan in 3.13 L 48% HBr and 6.26 L glacial acetic acid and heat on steam bath nine hours. Evaporate in vacuum to 3 L; add 3 L water; boil; decolorize; cool and filter. Extract filtrate with 2L $CHCl_3$ and evaporate in vacuum the extract. Add residue to precipitate. Filter out to get 3-Br-isoxazole-5-proprionic acid (I) (recrystallize-benzene-cyclohexane). Dissolve 44 g (I) in 440 ml concentrated sulfuric acid and add 80 g CrO_3 in 80 ml water dropwise over three hours at 15-20°. Pour onto 800 g ice and extract the aqueous solution with 3X500 ml ether. Evaporate in vacuum to get 3-Br-isoxazole-5-COOH (II) (re-crystallize-benzene-toluene). Dissolve 30.7 g (II) and 27 g KOH in 500 ml methanol and stir two hours at 140°. Cool, add 1.5 L water and extract with 3X1.5 L ether. Can decolorize aqueous solution by boiling briefly with carbon, then filtering and acidifying with concentrated HCl. Filter and recrystallize-benzene to get 3-methoxy-isoxazole-5-COOH, which is the analog of (IV) in method 1 and is identical to (II) in method 3 and can be converted to muscimole by either method.

Method 5 Rend. Ist. Lomb. 94A,737(1960)

This method gives a somewhat lower yield than method 1. Dissolve 14.1 g 1-nitro-3-butanone in 30 ml glacial acetic acid and heat to 35°; add slowly with stirring to a solution of 5.3 g bromine in 10 ml glacial acetic acid. Evaporate in vacuum, dissolve residue in ether and wash with water, $NaHCO_3$ and water. Dry and evaporate in vacuum the ether (can distill 84/2) to get 6 g 1-nitro-4-Br-3-butanone (I). Add 4.4 g (I) to 25 ml 48% HBr and reflux three hours. Add 50 ml water and steam distill. Neutralize the distillate with K carbonate and extract with ether. Dry and evaporate the extract (can distill 128/20) to get 5.6 g 2-Br-5-Br-methyl-isoxazole (II). To a warm solution of 5.2 g urotropine (hexamethylene-tetramine) in 45 ml $CHCl_3$, add 8 g (II) in 10 ml $CHCl_3$ and let stand four hours at room temperature. Filter, wash with $CHCl_3$ and dry. Dissolve 2.3 g precipitate in 10 ml concentrated HCl and let stand two hours at room temperature. Add 30 ml water, basify with NaOH and extract with ether. Precipitate with dry HCl gas to get 3-Br-5-aminomethyl-isoxazole (III). Test for activity. Dissolve 8.8 g (III) and 5.6 g KOH in 60 ml methanol and reflux thirty hours. Dissolve in 100 ml water, acidify with dilute HCl; evaporate in vacuum and dissolve residue in hot methanol to get 3-methoxy-5-aminomethyl-isoxazole (IV). Test for activity. Convert (IV) to muscimole as described in method 1 or as follows: reflux in concentrated HCl four hours and evaporate in vacuum (recrystallize-methanol:tetrahydrofuran 1:1). Can add tetraethylamine to precipitate.

Method 6 (best yield) CPB 19,46(1971), JCS 121,1638(1922)

To prepare diethylacetone dicarboxylate (I) proceed as in method 3 for cocaine synthesis, substituting 720 ml ethanol for methanol. To 197 g (I), add rapidly with stirring and cooling 208 g PCl_5, keeping temperature below 50°. When HCl evolution stops, pour into water and add ice as necessary to cool. Extract the red oil with ether and dry, evaporate in vacuum. Boil residue 2½ hours with 20% HCl, evaporate the water and dissolve the residue in ether. Dry and evaporate in vacuum to get 100 g 3-Cl-glutaconate (II). 100 g (II), 300 g ethanol, 50 ml sulfuric acid and reflux while passing in ethanol vapor until 1.5 L collects. Add water, extract with ether, wash with aqueous Na carbonate and dry, evaporate in vacuum to get 113 g diethyl-3-Cl-glutaconate (III). Mix 28 g NaOH, 140 ml water, 10.4 g hydroxylamine.HCl, 280 ml ethanol and cool to -35° in acetone-dry ice bath. Stir and add 30.3 g (III) in 50 ml ethanol over two minutes; stir one hour (can let stand twelve hours). Neutralize with

60 ml concentrated HCl while cooling in ice bath. Evaporate in vacuum and then add 62 ml dry ethanol, 130 ml benzene and ½ml concentrated sulfuric acid to residue and reflux twenty-four hours to remove 80 ml of the azeotrope. Cool, pour into 60 ml ice water and separate the organic layer. Extract the aqueous layer with 30 ml benzene and add to organic layer. Wash organic layer with 2X15 ml NaHCO$_3$ (2%), 2X30 ml water and dry, evaporate in vacuum to get 14 g ethyl-3-OH-5-isoxazoleacetate (IV) (recrystallize-ether). Add 3.36 g (IV) portionwise to 1.35 g hydrazine hydrate which has been heated at 94° on water bath and heat 3 hours. Cool, filter and recrystallize-methanol to get 2.6 g of the hydrazide (V) (concentrate filtrate to get more). Dissolve 3.41 g (V) in 30 ml tetrahydrofuran and mix with 6.55 ml of a solution of dry HCl in tetrahydrofuran (9.94M). Cool to -50° in dry ice-acetone bath and add dropwise 2.81 g isoamyl nitrite in 10 ml tetrahydrofuran over five minutes. Remove cooling bath, let stand fifteen minutes and stir fifteen minutes and filter. Evaporate in vacuum the liquid, dissolve the residue in 80 ml ethyl acetate; wash with 2X20 ml 1:9 mixture of saturated aqueous NaHCO$_3$ and saturated aqueous NaCl and dry, evaporate in vacuum. Dissolve oily residue in 40 ml dry ethanol and reflux on oil bath forty-five minutes. Cool, evaporate in vacuum, filter and wash with benzene to get 1.2 g 3-OH-5-(carbethoxyamino) methyl-isoxazole (VI) (recrystallize-benzene). 3.83 g (VI) in 80 ml 20% Ba(OH)$_2$. Reflux fifteen hours and evaporate in vacuum (or cool and acidify with 10% sulfuric acid, filter, and evaporate in vacuum) to get muscimole (recrystallize-methanol).

Other References

The following papers describe the preparation of various muscimole analogs which may have activity:

CJC 48,3753(1970); TL 2101(1970); JOC 35,1806(1970); JMC 13,738(1970); BSC 2685(1970). For a simple two-step synthesis of 3-OH-5-methyl-isoxazole (which may substitute for (III) in method 1) see CA 74,76407(1971) (cf. CA 74,125521(1971)). The antibiotic Oxamycin (cycloserine) is often psychedelic (see Merck Index for synthesis).

LSD

Since Hofmann's first trip in 1943, great deal of interest has been generated in the occurrence and properties of various lysergic acid derivatives. Fungi of the genus *Claviceps,* which grow on rye wheat, rice and other grasses, were the first natural source of these alkaloids to be discovered. In recent years related compounds have been found in the genera *Penicillium* (the blue-green mold that also produces penicillin), *Aspergillus,* and *Rhizopus* (the black bread mold). These compounds are now produced commercially by culturing certain strains of *Claviceps* which produce as much as 4 g of ergotamine per Liter of culture medium. Growing pure cultures of fungi is not for amateurs, but those interested will find these references useful: JPS 58,143(1969); App. Microbiol. 18,464 (1969); HCA 47,1052(1964); Lloydia 32,327,401(1969); Can. J. Microbiol. 16,923(1970); CA 61,15314c-f, 67,84858e, 69, 36323w; Biotech. Bioeng. 13,331(1971); CA 76,57736(1972); U.S. Patent 3,483,086; Planta Med. 23,330(1973); J. Pharm. Educ. 36,598(1972); CA 78,41492(1973); French Patent 1,531, 205; German Patents 1,806,984 and 1,909,216; British Patent 1,158,380.

For a description of a wild American Claviceps species see Mycologia 66,978(1974).

The occurrence of hallucinogens in the seeds (and to a lesser degree in the leaves and stems) of various members of the family Convolvulaceae (morning glories, etc.) was known to the Aztecs. Seeds of the genera *Rivea, Impomoea,* and *Argyria* (Hawaiian baby woodrose) contain lysergic acid derivatives; the woodrose being champion with about one hundred times as much as the other genera (about 7 mg alkaloids/g seeds). In view of the low yield (maximum 10 mg alkaloids/100 g seeds) even the famed pearly gates variety of morning glory is not worthwhile extracting, and the trip is commonly a bummer, resembling that produced by scopolamine or ibogaline and unlike that of LSD. However, the lysergic acid amide, etc., can be extracted, hydrolyzed to lysergic acid (as described below for ergot alkaloid hydrolysis), and converted to LSD by any of the methods described. For species variation of alkaloid content see Lloydia 29,35(1966). Crude ergot or woodrose seeds should yield ca. 1 g LSD/kg after conversion of the isolated alkaloids.

Alkaloid Extraction (short method)

Finely grind seeds (preferably woodrose) and add $NaHCO_3$. Extract with ethyl acetate by soaking about one day. Filter and extract the ethyl acetate with tartaric acid solution. Basify the extract with $NaHCO_3$ and extract it with ethyl acetate. Dry and evaporate in vacuum the ethyl acetate to get the alkaloids. Repeat this procedure on the seeds until no more residue is obtained.

Alternatively, add 100 ml petroleum ether to 100 g finely ground seeds and let soak about two days. Filter, discard petroleum ether and let seeds dry. Add 100 ml methanol to the seeds and let soak about two days. Filter, repeat extraction with another 100 ml methanol and evaporate in vacuum the combined methanol extracts. The residual yellow oil contains the alkaloids.

For chromatographic purification of ergot alkaloids from seed extracts see Phytochem. 11,1479(1972).

For ergot extraction and separation see also Fr. Patent 2,089,081 (11 Feb 1972) and CA 79,105,457(1973).

For a recent review of the ergot alkaloids see R. Manske (Ed.) The Alkaloids, vol. 15:1-40(1975).

Extraction of Lysergic Acid Amides from Woodrose Seeds or Powdered Ergot

Reduce the seed material to a fine powder in a blender, and spread it out to dry. Grind it again if it is not fine enough after the first time due to dampness.

Saturate the powdered seed material with lighter fluid, naphtha or ligroine. When completely saturated, it should have the consistency of soup.

Pour it in a chromatography column and let it sit overnight.

Remove the fatty oils from the material by dripping the lighter fluid or other solvent through the column slowly and keep testing the liquid that comes through for fats by evaporating a drop on clean glass until it leaves no greasy film. It will take several ounces of solvent for each ounce of seeds.

Mix 9 volumes of chloroform with 1 volume of concentrated ammonium hydroxide and shake it in a separatory funnel.

When it settles the chloroform layer will be on the bottom. Drain off the chloroform layer. Discard the top layer.

Drip the chloroform wash through the column and save the extract. Test continuously by evaporating a drop on clean glass until it ceases to flouresce under a black light.

Evaporate the chloroform extracts and dissolve the residue in the minimum amount of a 3% tartaric acid solution. If all the residue doesn't dissolve, place it into suspension by shaking vigorously.

Transfer the solution to a separatory funnel and wash the other vessel with acid in order to get all the alkaloid out. Pour the washings in the funnel also.

Basify by adding sodium bicarbonate solution, and add an equal volume of chloroform.

Shake this thoroughly, let it settle, remove the bottom layer and set it aside.

Once again, add an equal portion of chloroform, shake, let it settle and remove the bottom layer.

Combine the chloroform extracts (bottom layers) and evaporate to get the amides.

The Culture and Extraction of Ergot Alkaloids

Make up a culture medium by combining the following ingredients in about 500 milliliters of distilled water in a 2 liter, small-neck flask:

Sucrose	100 grams
Chick pea meal	50 grams
Calcium nitrate	1 gram
Monopotassium phosphate	0.25 grams
Magnesium sulphate	0.25 grams
Potassium chloride	0.125 grams
Ferrous sulphate heptahydrate	8.34 milligrams
Zinc sulphate heptahydrate	3.44 milligrams

Add water to make up one liter, adjust to pH 4 with ammonia solution and citric acid. Sterilize by autoclaving.

Inoculate the sterilized medium with *Claviceps purpurea* under sterile conditions, stopper with sterilized cotton and incubate for two weeks periodically testing and maintaining pH 4. After two weeks a surface culture will be seen on the medium. Large-scale production of the fungus can now begin.

Obtain several ordinary 1 gallon jugs. Place a two-hole stopper in the necks of the jugs. Fit a short (6 inch) glass tube in one hole, leaving 2 inches above the stopper. Fit a short rubber tube to this. Fill a small (500 milliliter) Erlenmeyer flask with a dilute solution of sodium hypochlorite, and extend a glass tube from the rubber tube so the end is immersed in the hypochlorite. Fit a long, glass tube in the other stopper hole. It must reach near the bottom of the jug and have about two inches showing above the stopper. Attach a rubber tube to the glass tube as short or as long as desired, and fit a short glass tube to the end of the rubber tube. Fill a large, glass tube (1 inch x 6 inches) with sterile cotton and fit 1-hole stoppers in the ends. Fit the small, glass tube in end of the rubber tube into 1 stopper of the large tube. Fit another small glass tube in the other stopper. A rubber tube is connected to this and attached to a small air pump obtained from a tropical fish supply store. You now have a set-up for pumping air from the pump, through the cotton filter, down the long glass tube in the jug, through the solution to the air space in the top of the jug, through the short glass tube, down to the bottom of the Erlenmeyer flask and up through the sodium hypochlorite solution into the atmosphere. With this aeration equipment you can assure a supply of clean air to the *Claviceps purpurea* fungus while maintaining a sterile atmosphere inside the solution.

Dismantle the aerators. Place all the glass tubes, rubber tubes, stoppers and cotton in a paper bag, seal tight with wire staples and sterilize in an autoclave.

Fill the 1-gallon jugs 2/3 to 3/4 full with the culture medium and autoclave.

While these things are being sterilized, homogenize in a blender the culture already obtained and use it to inoculate the media in the gallon jugs. The blender must be sterile. Everything must be sterile.

Assemble the aerators. Start the pumps. A slow bubbling in each jug will provide enough oxygen to the cultures. A single pump can, of course, be connected to several filters.

Let everything sit at room temperature (25° C.) in a fairly dark place (never expose ergot alkaloids to bright light -- they decompose) for a period of ten days.

After ten days adjust the cultures to 1% ethanol using 95% ethanol under sterile conditions. Maintain growth for another two weeks.

After a total of 24 days growth period the culture should be considered mature. Make the culture acidic with tartaric acid and homogenize in a blender for one hour.

Adjust to pH 9 with ammonium hydroxide and extract with benzene or chloroform/iso-butanol mixture.

Extract again with alcoholic tartaric acid and evaporate in a vacuum to dryness. The dry material is the salt (i.e., the tartaric acid salt, the tartrate) of the ergot alkaloids, and is stored in this form because the free basic material is too unstable and decomposes readily in the presence of light, heat, moisture and air.

To recover the free base for extraction of the amide or synthesis to LSD, make the tartrate basic with ammonia to pH 9, extract with chloroform and evaporate in vacuo.

If no source of pure *Claviceps purpurea* fungus can be found, it may be necessary to make a field trip to obtain the ergot growths from rye or other cereal grasses. Rye grass is by far the best choice. The ergot will appear as a blackish growth on the tops of the rye where the seeds are. They are approximately the same shape as the seeds and are referred to as "heads of ergot." From these heads of ergot sprout the *Claviceps purpurea* fungi. They have long stems with bulbous heads when seen under a strong glass or microscope. It is these that must be removed from the ergot, free from contamination, and used to inoculate the culture media. The need for absolute sterility cannot be overstressed. Consult any elementary text on bacteriology for the correct equipment and procedures. Avoid prolonged contact with ergot compounds, as they are poisonous and can be fatal.

LSD Identification

Since LSD is an indole derivative, it gives a positive reaction (violet color) to the tests given in the indole section. LSD also fluoresces under an ultraviolet light (black light), but so do many other compounds. For infrared spectra of LSD and related compounds, see JACS 78,3087(1956) and J. Forensic Sci. 12,538 (1967). For other information on identification see JPS 56,1526 (1967) and JAOAC 50,1362(1967), 51,1318(1968). For a microcrystalloscopic test see J. Pharm. Pharmacol. 22,839(1970).

In order to make LSD, lysergic acid is needed. This can sometimes be obtained, but generally one of the lysergic acid containing ergot alkaloids such as ergotamine is more readily available. Ergot is the dried sclerotium of various species of fungi which infect rye (and other grasses), leading to the formation of large purple growths in place of the rye grains. These growths are collected, dried, powdered and the alkaloids extracted. For the

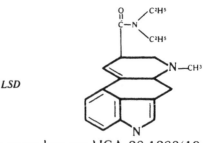

LSD

extraction procedure see HCA 28,1283(1945), J. Pharm. Pharmacol. 7,1(1955), JPS 50,201(1961), CA 75,137422(1971). Proc. Indian Acad. Sci. 71B,28,33(1970) gives production from artificially infected rye. Ergot is produced mainly in Europe (especially Switzerland) but some has been grown in the USA (e.g., in Minnesota). This production occurs primarily because of the use of ergotamine and related compounds in medicine (contracting the post-partum uterus, terminating migraine headaches, etc.). Many of the ergot alkaloids are derivatives (amides) of lysergic acid. Unfortunately, these compounds have little hallucinogenic activity and it is necessary to hydrolyze (split with water) off the amide, producing lysergic acid, and to synthesize a different amide with greater psychedelic activity. This hydrolysis can be done with any of the following compounds or a mixture of them: ergometrine, ergine, ergotamine, ergosine, ergocristine, ergokryptine, ergonovine (ergometrine) and methysergide (Sansert). When -ine is added to the name (e.g., ergotaminine) this indicates the isomers which will lead to the production of the inactive iso-LSD. The papers cited here give simple techniques for converting these to the active forms (or see the technique for converting iso-LSD to LSD in method 1 following): HCA 37,820,2039(1954); CA 69,36322(1968); CCCC 34,694 (1969). For a review of the ergot alkaloids see THE ALKALOIDS, Manske and Holmes (Eds.), 8,725(1965), and F. Bove, THE STORY OF ERGOT (1970).

Ergot Alkaloid Hydrolysis

JBC 104,549(1934); HCA 47,1929(1964). Perhaps the best method is Hofmann's modern hydrazine hydrolysis given later, since this disposes of the necessity for isolating the lysergic acid (I); otherwise the following alkaline hydrolysis can be used:

Dissolve 20 g of the alkaloid (e.g., ergotamine) in 200 ml 1M KOH in methanol (i.e., dissolve 56 g KOH pellets in 1L 100% methanol) in a 1 L heavy walled vacuum flask and evaporate in vacuum the methanol at room temperature. To prevent the solution from cooling, and thus greatly prolonging the evaporation time, put the flask in a pan of water kept at room temperature by gentle heating or by running warm water through it. Add 400 ml 8% KOH in water to the residue and boil for one hour (under N_2 if possible -- this can be done by filling the flask with a N_2 stream and loosely stoppering or by allowing a gentle stream of N_2 to flow through during heating). Cool, acidify with dilute sulfuric acid and shake in separatory funnel with 1 L ether. Discard the upper ether layer and filter with vacuum the aqueous suspension of lysergic acid (I). Wash precipitate with 20 ml dilute sulfuric acid. To recover the small amount of (I) remaining in solution, basify with Na carbonate and bubble CO_2 through it. Filter and add precipitate to first batch. Some isolysergic acid will remain in solution and can be precipitated by adding 10% HNO_3. It can be converted to (I) by adding 3 ml 10% KOH for each 0.1 g acid, boiling on steam bath for one hour under N_2 (if possible) and precipitating by acidifying with glacial acetic acid. Maximum yield is about 9 g (I) for 20 g ergotamine.

A shorter method of hydrolysis which may work as well follows: dissolve 20 g alkaloid in 300 ml methanol and 300 ml 40% KOH and reflux two hours under N_2 (if possible). Cool, saturate with CO_2 and evaporate in vacuum. Extract the residue with hot ethanol three times and dry, evaporate in vacuum the combined ethanol extracts to get (I).

Under ordinary conditions, about 20% of (I) will be converted by the action of hot water, etc., to the inactive isolysergic acid. Most of this remains in solution and can be isomerized to (I) as described above, or it can be converted to iso-LSD by any of the methods described later and isomerized to LSD (see method 1).

It is unnecessary to purify (I), but this can be done as follows: dissolve 9 g (I) in 20 ml NH_4OH, filter and concentrate in vacuum at room temperature to precipitate (I). After filtering, the grey crystals can be further purified by dissolving in boiling water and cooling in ice bath to precipitate (I). Melting point should be about 240° (decomposes). Alternatively, the dark-colored (I) resulting from hydrolysis can be shaken with 2X400 ml 2 M NH_4OH in ethanol, and the combined extracts evaporated in vacuum to give (I). Dissolve the remaining residue in 500 ml hot methanol, cool to 0° and filter out the (I) (recrystallize-water). Can remove colored impurities by shaking solution with decolorizing carbon and filtering.

Recently a method for increasing the yield of (I) about 10% using 2.5% hydrazine hydrate was described (CA 69,36323w(1968)).

Dissolve 7 g alkaloid in 200 ml 6 N KOH in methanol and 200 ml ethanol, add 10 ml hydrazine hydrate and boil four hours under N_2 (if possible) and proceed as above.

Finally, the (I) must be thoroughly dried by heating at about 110°/1 mm for two hours or 150° if ordinary lab vacuum of 15 mm is used. A forced water vacuum (about 25 mm) can be used here as elsewhere. An oil bath (e.g., mineral oil) will allow temperature regulation.

LSD Synthesis
Dangers

There are certain aspects of LSD production which are common to all synthetic methods. The first is a certain degree of danger; each uses dangerous reagents and solvents.

Hydrazine and hydrazine hydrate are both violent poisons, and each can cause severe skin burns and eye damage. The vapor of each is irritating, and can cause severe eye irritation as well as liver and blood damage, but the symptoms don't always manifest right away, sometimes appearing three or four days after exposure, so it is easy for exposure to be much more dangerous than is immediately realized. In addition, anhydrous hydrazine is a sensitive and violent explosive, the explosion of which can be set off by certain types of stainless steel and such common things as wood and rust.

Both triflouroacetic acid and sulfur trioxide will cause very severe skin burns, and their vapors are extremely irritating. Sulfur trioxide is such a strong dehydrating agent that it chars organic material, and its heat of dehydration is so high that it will start a fire if spilled on wood, which could prove fatal were flammable solvents in use at the time or stored nearby.

Phosgene is very poisonous; so insidious that it was used as a war gas in World War I. One deep breath can cause immediate collapse and death, and as it is not irritating there is no gag reflex to prevent one from taking that deep breath. Doses which are not high enough to be immediately lethal may not be noticed at all at the time of exposure, yet lead to death within 24 hours. Sub-lethal doses cause pulmonary edema and serious respiratory disability; again, the symptoms can appear well after an exposure which was hardly noticed.

Diethylamine, used in every LSD synthesis, has a very low flash point, and its vapor is irritating. The vapor of DMF is also irritating, and prolonged exposure can cause liver damage. In fact, most of the solvents used in LSD production are either flammable or toxic or both.

In addition to all the above, the starting material, the ergot alkaloids, is as a class quite toxic, and clean working conditions are necessary when working with it. Ergot alkaloid poisoning, known in the Middle Ages as Saint Antony's fire, can actually cause one's limbs to blacken, shrivel, and fall off! Any woman working with these compounds should also be aware that many of them are oxytoxics, that is, they cause uterine contractions, and are so used to induce labor, etc.

Working Conditions

There are certain procedures common to all syntheses of LSD which are based upon the sensitive nature of ergot compounds in general. Natural ergot alkaloids, lysergic acid, LSD, and the intermediate products associated with the various syntheses are all to a varying degree unstable. Even the most stable of these compounds will readily decompose under any but moderate conditions. Thus precautions must be taken against light, moisture, oxygen, and heat.

Light of the ultraviolet region promotes addition of water at the $\Delta 9$-10 double bond to form the lumi-compounds. Thus reactions are best carried out in the light of red or yellow photographic darkroom bulbs, and storage should be in opaque or amber bottles.

Most of the reactions involved in LSD synthesis require anhydrous conditions for good yield, and so protection must be made against moisture during the actual production. Furthermore, the final product must be thoroughly dried to prevent possible formation during storage of the lumi-compounds as mentioned above.

Oxidizing agents, including atmospheric oxygen, will decompose ergot compounds. For this reason, all reactions are carried out in an atmosphere of an inert gas such as nitrogen. The danger of oxidation increases with temperature, so this precaution is of course most important with those reactions proceeding at elevated temperature.

Various methods have been devised to prevent oxidation during storage. The most obvious is to store the LSD in nitrogen filled containers, but the excellent protection thus afforded is of course lost when the bottle or ampule is opened. Another method is to use an antioxidant; Brown and Smith recommend ascorbic acid. A more sophisticated method, recommended on the highest authority, is to make LSD maleate rather than the tartrate. Both maleic and tartaric acids are dicarboxylic, but the pK^2 of maleic acid is too low to form a salt with LSD. Thus there is a free carboxyl group in LSD maleate, which group will serve to prevent oxidative decomposition.

Excessive heat will cause decomposition of LSD and its precursors, and will also increase the possibility of racemization. Thus reactions at elevated temperature are not unnecessarily prolonged, nor are temperatures unnecessarily raised. All drying is done in vacuo in an inert atmosphere, and long term storage should be under refrigeration.

Legal Acid

I want to emphasize that "legal acid" can be obtained if other amines are substituted for diethylamine in LSD synthesis. These other lysergamides should give identical trips, but most of them are less potent than LSD. Precise potency data do not exist, so it remains for an enterprising chemist to gain immortality by adding each of the following amines (and any others that come to mind) to separate aliquots of the final step of LSD synthesis (they could easily be done simultaneously), isolating the tartrates and assaying them for potency: piperidine, diisopropylamine, ethylisopropylamine, ethylpropylamine, methylethylamine, methylisopropylamine, tetra-hydrooxythiazine, tetrahydroisoxazine, dioxazole, 2-methylmorpholine, 2,5-dimethyl (or dimethoxy) pyrrolidine, cyclo-butyl-amine, cyclopentylamine, etc. Published potency data expressed as a fraction of LSD activity follow: pyrrolidide (1/20), dimethylamide (1/20), morpholide (1/10 or 1/3), ethylpropyl (1/3), dipropyl (1/10), methylethyl (less than 1/10), methylpropyl (less than 1/10).

Methods

LSD via the Hydrazide HCA 38,429(1955), HCA 26,953(1943) CA 57,12568(1962), U.S. Patent 3,239,530(1966).

Perhaps the simplest method is the following devised by Hofmann, which proceeds directly from the ergot alkaloid via hydrazine, and is not to be confused with his earlier use of hydrazine under more violent conditions which led to a racemized product and reduction of the yield by one-half.

Add 1.16 g ergotamine.HCl to 4 ml anhydrous hydrazine and heat one hour at 90°. Add 20 ml water and evaporate in vacuum. Can proceed to the next step or can purify by adding ether and aqueous tartaric acid, basify the aqueous phase and extract aqueous phase with $CHCl_3$ to get mainly d-iso-lysergic hydrazide (I). Can chromatograph on alumina and elute with 0.5% ethanol in $CHCl_3$ to purify. To 1 g (I), finely ground, in 40 ml 0.1 N ice cold HCl, add with good stirring at 0° 4 ml 1N Na nitrite. Quickly over two-three minutes, add 40 ml 0.1 N HCl so pH is about 5. Let stand five minutes, basify with 1N $NaHCO_3$ and extract with 100 ml ether, then 50 ml ether. Wash ether with water and dry and evaporate in vacuum at 10°. Dissolve the resulting yellow azide in about 5 ml diethylamine (DEA) at 0° and heat one hour at 60° in a bomb (sealed metal pipe), or heat 3 to 4 hours at 45° C in a vented flask. Let stand several hours and evaporate in vacuum to get about 0.7 g d-LSD and 0.15 g d-isoLSD (which can be converted to d-LSD as described in method 1 following). Alternatively, the DEA can be added to the cooled ether solution of the azide and let stand several hours or overnight at room temperature in the dark in a vented flask.

An alternate method of proceeding from the hydrazide follows (U.S. Patent 3,085,092). To a solution of 1.4 g (I) in 5.5 ml 1 N HCl, 5 ml water, 5 ml EtOH, add 1 g acetylacetone (2,4 pentanedione), let stand 1 hour at room temperature and neutralize with 5.5 ml 1 N NaOH. Filter off the lysergyl pyrazole (II) and wash with water. Can purify by drying in vacuum at 60° C and recrystallizing from chloroform by the addition of ether. Heat 0.4 g (II) and 2.5 ml DEA at 100° C for 2 hours (or let stand 15 hours at room temperature, evaporate to dryness and heat a few minutes at 100° C in vacuum). Can recrystallize from $CHCl_3$, petroleum ether or as described elsewhere here.

Technical Scale Details For This Method
Hydrazide Production

In dim yellow light, (preferably) three tared and fully dried 250 ml round-bottom flasks containing stirring bars are each charged with 30 g dry ergotamine tartrate and 120 ml anhydrous hydrazine. The flasks are fitted with gas inlet tubes adjusted to just above the liquid level and streams of nitrogen passed through, the exhaust gas being led through wash bottles equipped with traps and containing dilute acid to remove hydrazine vapors. The flasks are lowered into oil baths preheated to 90°, and heated with slow stirring for one hour. The contents of the three reaction flasks are then emptied into a 2000 ml beaker containing 900 ml distilled water, and this solution transferred to a 3000 ml two-neck round bottom flask. An additional 900 ml water is used to rinse the residue in the flasks, beaker, etc. into the 3000 ml flask. This large flask is fitted with siphon tube, gas inlet tube, and gas outlet connected to wash bottle and trap.

The aqueous hydrazide solution is evaporated from a tared 2000 ml flask on an efficient rotary evaporator, using a bath temperature of 40° and an ice-cooled condenser; the 3000 ml siphon flask assembly is used as storage for the vacuum feed. The weight of the crude hydrazide so obtained is determined, it is dissolved in about 170 ml 1 N tartaric acid, the aqueous solution washed with three 30 ml portions ether, made alkaline with 190 ml 1 N ammonium hydroxide, and exhaustively extracted with successive portions of chloroform, the first two portions being 100 ml each, the following 50 ml.

Completeness is ensured by testing with UV light, extraction ceasing only when the chloroform extract exhibits no blue fluorescence. The chloroform solution is washed with three 30 ml portions distilled water, dried over chloroform moistened magnesium sulfate, and the hydrazide recovered by vacuum evaporation in tared 500 ml flasks, one such flask being used for each two 90 g batches. These flasks are flushed with nitrogen, stoppered, and stored in a dark and dry refrigerator. As the hydrazide is stable, all the ergotamine tartrate will be converted to it prior to the next step. Theoretical yield from 1000 g ergotamine tartrate is 429.65 g; 80% yield is 343 g.

Pyrazole Production

In dim red light, the weighed hydrazide contained in one of the 500 ml flasks (ca. 67 g; 95% of theory) is washed into a 1000 ml beaker with 263 ml 1N hydrochloric acid. 239 ml distilled water, 239 ml ethanol (95%), and 37 ml 2,4-pentanedione are added, and the well-mixed solution left to stand in the dark at room temperature until the reaction is complete, i.e., about 30 minutes. The reaction mixture is neutralized with the addition of 263 ml 1 N sodium hydroxide, and the beaker covered with parafilm and refrigerated to ensure complete precipitation. The pyrazole is filtered at the pump, the mother liquor being returned to the beaker and used to wash out the last few crystals, washed with cold water, and sucked dry under a stream of dry nitrogen. The product is dried in vacuo over barium oxide or phosphorus pentoxide for at least twelve hours before proceeding to the next step, wherein anhydrous conditions will increase yield. Hofmann calls for drying the pyrazole in vacuo at 60°, which indicated the product to be fairly stable. So all the hydrazide is converted prior to aminization.

Amide Production

In dim red light, 50 g of the well-dried pyrazole and 700 ml freshly dried diethylamine are placed in a tared and well-dried 1500 ml flask equipped with gas inlet tube and stirring bar. The flask is lowered into a bath preheated to 45°, and the contents stirred under a stream of nitrogen for four hours. On a rotary evaporator, using a bath temperature of 40° and an ice-cooled condenser, the diethylamine is removed in vacuo and set aside for purification and re-use. Briefly and in high vacuum the flask is heated to 100°, the split-off pyrazole being thereby driven off.

The residue so obtained is immediately placed in solution with methanolic potassium hydroxide to effect interconversion of the stereoisomers. Amination and Transposition will proceed simultaneously, the first batch being transposed while the second is aminated.

Production Scale Isomerization of iso-LSD to LSD

In dim red light, the amide residue from the last step is dissolved in the least possible amount of dry methanol and washed into a 1500 ml round-bottom flask. A two-fold volume of 4 N methanolic potassium hydroxide is added, and the well-mixed solution left to stand at room temperature, in the dark and under a slow stream of nitrogen, for four hours. At the end of this period, the solution is neutralized with methanolic hydrogen chloride (ca. 5 N), washed into a 4000 ml Erlenmeyer flask, and dried over methanol-moistened anhydrous magnesium sulfate (0.10 g $MgSO_4$ per ml KOH solution). The methanolic acid should be added slowly and with good stirring to prevent possible hydration of the 9-10 double bond to give lumi-LSD. Together with 100 ml dry isopropanol (to remove the last trace of water azeotropicly) the dried solution is transfered to a 3000 ml siphon flask assembly, and the solvent removed in vacuo in a tared 500 ml two-part freeze-dry flask. The weighed gummy residue is scraped into the thimble of a Soxhlet extractor, the adhering residue being washed into the thimble with portions of warm chloroform, the total volume of which is 12.5 ml per gram amide (total weight minus weight KCl). A 3000 ml flask is used with the extracter, and it is previously charged with 37.5 ml dry benzene per gram amide. Under a stream of dry nitrogen, the solvent is in vacuo at 40° refluxed through the thimble, thus extracting the amide from the inorganic salt and at the same time preparing the solution for use in the chromatographic separation of the stereoisomers. The above solution is stored over a small amount of benzene-moistened calcium sulfate in a nitrogen flushed flask which is placed in a dark refrigerator. All the pyrazole is converted to this benzene-chloroform solution prior to separation of the isomers.

The following methods all proceed from lysergic acid (I). Methods 1, 2, 4, and 6 give less than 20% iso-LSD in the product but methods 2, 5, and 9 seem to have the highest total yield (about 80%) of LSD plus iso-LSD. Since unreacted lysergic acid can be recovered and run through the synthesis again, and iso-LSD isomerized to LSD as described here, it is probably best to use the simplest methods. These comparative yields come mostly from the reference to method 9.

From Lysergic Acid — Method 1 CA 50,10803d(1956) (Pioch)

Dissolve 5.3 g dry (I) in 125 ml acetonitrile (or dimethyl-formamide or proprionitrile) and cool to -20° (freezer or dry ice-acetone or ethanol mixture). Add 8.82 g trifluoroacetic anhydride in 75 ml acetonitrile cooled to -20° carefully. Let stand at -20° 1½ hours or until all the (I) dissolves. Then add 7.6 g DEA in 150 ml aceatonitrile and let stand at room temperature in dark two hours. Evaporate in vacuum to get LSD. If purification is desired, dissolve the residue in 150 ml CHCl$_3$ and add 20 ml ice water. Pour into ½ L separatory funnel and drain out the lower CHCl$_3$ layer into a beaker (after shaking). Add 50 ml CHCl$_3$ to funnel, shake and drain bottom layer into same beaker. Repeat with 3X50 ml CHCl$_3$ and discard the water. Extract the combined CHCl$_3$ extracts with 4X50 ml ice cold water and dry, evaporate in vacuum the CHCl$_3$ to get 3.5 g d-LSD. This is composed partly of the inactive d-iso-LSD, which, while it will not effect the trip, can be recovered and converted to d-LSD as follows: dissolve the residue in 120 ml benzene and 40 ml CHCl$_3$ (or 200 ml methanol), add tartaric or maleic acid and shake to precipitate mainly d-LSD (add a little ether and cool in refrigerator several days if necessary to ensure complete precipitation; evaporate in vacuum the solvent to get d-iso-LSD. Add 50 ml ethanol and 5 ml 4N KOH per g iso-LSD and let stand at room temperature two hours; evaporate in vacuum (or extract with CHCl$_3$ as above) to get LSD. 4 hours in 2.66 N methanolic KOH is said to be optimal for isomerization.

From Lysergic Acid — Method 2 JOC 24,368(1959) (Galbrecht)

This method is supposed to give little iso-LSD but it gives some of the monomethylamide.

Add 1 L dimethylformamide (freshly distilled, if possible) to dry flask fitted with stirrer, ice bath, dropping funnel and condenser, both protected from water by Ca chloride drying tubes. Add dropwise with stirring over four-five hours at 0° 0.21 lbs (90.7 g) SO_3 (sulfuric anhydride, available as Sulfan from Allied Chem. Co.). If precipitate forms, stir until it dissolves. Sulfan may be made in larger amounts and is good for several months if kept dry and cool. Molarity of fresh SO_3-DMF reagent should be about 1M, but for precise determination, add a little water to an aliquot and titrate with standard NaOH to phenolphthalein end point. Add 6.45 g dry (I) (or 7.15 g (I) monohydrate) and 1.06 g LiOH hydrate (or NaOH or KOH but these absorb water so they must be dissolved in absolute methanol, titrated and added in equimolar amounts) to 200 ml methanol in a 1 L vacuum flask and evaporate in vacuum. Dissolve residue in 400 ml anhydrous dimethylformamide and distill off 200 ml DMF at about 15 mm Hg through a twelve inch column packed with glass helices or other material. Cool to 0° and rapidly add 50 ml SO_3-DMF solution (1 M). Stir at 0° for ten minutes and add 91.5 g (12.9 ml) DEA and stir ten minutes. Add 400 ml of water, stir and add 200 ml saturated NaCl. Extract the LSD by shaking with several 500 ml portions ethylene dichloride (can use indole test given in indole section to show completeness of extraction). Combine extracts (lower layer in separatory funnel) and dry, evaporate in vacuum to get LSD (can purify as above).

From Lysergic Acid — Method 3 JOC 24,368(1959) (Garbrecht)

This route is said to give a lower yield than method 2. Dissolve 13.4 g dry (I) in 250 ml dry dimethylformamide and cool to 0°. Add cooled solution of 3.4 ml 0.35 M methane-sulfonic acid anhydride in dry dimethylformamide. After thirty minutes at 0° add 14.6 g (20.4 ml) DEA and keep at 0° one hour. Evaporate in vacuum to get LSD and proceed as above.

From Lysergic Acid — Method 4 CA 57,5979(1962) (Hofmann)

Dissolve 0.536 g (I) in 10 ml freshly distilled $POCl_3$; stir and add 416 mg powdered, freshly sublimed PCl_5. Hold two minutes at room temperature, two minutes at 90°, and evaporate in vacuum. Extract the residue with hexane to give the lysergic acid chloride-HCl (can also extract the reaction mixture with hexane instead of evaporating in vacuum). Alternatively, use 6 ml $POCl_3$ and 240 mg $SOCl_2$ and heat three minutes at 90° to get the acid chloride. To 5 g of the acid chloride add 1.4 ml DEA in 50 ml methylene chloride and cool to 0°. Stir and add 27.5 ml pyridine and stir one-half hour at 0°. Warm to room temperature and stir 1½ hours; evaporate in vacuum to get LSD.

From Lysergic Acid — Method 5 CCCC 27,1590(1962) (Cerny

and Semonsky) cf. CA 75,77110(1971)

To a suspension of 13.4 g dry (I) in 800 ml dry dimethyl-formamide (DMF) in a 2 L vacuum flask at 20°, add a solution of 8.9 g N,N'-carbonyldiimidazole in 250 ml DMF and stir at 20° in dark for one-half hour. Add a solution of 4 g DEA in 50 ml DMF and let stand two hours at 20°; then twenty hours at 5°. Evaporate in vacuum to get LSD. Can purify as above or dissolve residue in 2½ L 2% tartaric acid; extract with ether and discard ether. Filter, basify with NH$_4$OH and extract with a 9:1 solution of ether:ethanol. Dry and evaporate in vacuum to get LSD in 81% yield.

From Lysergic Acid — Method 6 JMC 16,532(1973) (Johnson et al.)

This method gives very little iso-LSD.

To a refluxing slurry of 3.15 g dry (I) or monohydrate) in 150 ml CHCl$_3$ add 0.1 mole of the amine in 25 ml CHCl$_3$ and 2 ml POCl$_3$ simultaneously from separate dropping funnels over 2 to 3 minutes. Reflux 3 to 5 minutes more til a clear amber solution results. Cool to room temperature and wash with 200 ml 1M NH$_4$OH. Dry and evaporate in vacuum (below 40°). Can dissolve in the minimum amount of methanol and acidify with a fresh solution of 20% maleic acid in methanol. Filter and wash crystals with cold methanol to get the LSD or other amide. This method works with a wide variety of amines. For LSD itself, the POCl$_3$ can be added first. The yield is about 50%.

From Lysergic Acid — Method 7 German Patent 1.965,896 (1970) (Julia et al.)

See end of total synthesis of LSD given below.

From Lysergic Acid — Method 8 U.S. Patent 3,141,887 (Patelli and Bernardi)

Note: Phosgene is *very hazardous* and only professional chemists working with a fume hood should even think about using this method. It can be dissolved in a weighed container of DMF (dimethylformamide) and a second weighing will give the phosgene concentration.

0.5 g anhydrous lysergic acid suspended in 10 ml DMF or acetonitrile at -10° C are reacted with 2 ml of DMF containing 0.34 g of the phosgene-DMF complex for 20 minutes. Add 0.7 g diethylamine in 10 ml DMF (or acetonitrile), keep at -10° C for ten minutes and then at room temperature for 10 minutes. Dilute with chloroform, wash with NaOH (1 normal), then water and distill off the solvent in a vacuum. Dissolve the oily residue in methanol, acidify with tartaric or maleic acid; add ether to start crystallization. Keep overnight at 0° C, filter and wash with ether. Dissolve the product in methanol, decolorize with charcoal and precipitate with ether to obtain the tartrate or maleate (tartrate melts 192-198° C). $D^{20} = + 25°$ (C = 1 in water).

From Lysergic Acid — Method 9 CT 13,373(1978) (Losse and Mahlberg)

This method gives about 2/3 iso-LSD, and about 80% total yield.

Add 134 mg (0.5 mM) dry (I), 103 mg (0.5mM) N,N-dicyclohexylcarbodimide, 90 mg (0.67mM) 1-hydroxybenzotriazol (N-hydroxybenzotriazol) and 0.5mM diethylamine to 2.5 ml CH_2Cl_2 and 2.5 ml tetrahydrofuran and stir in the dark at 20° C for 24 hours. filter, wash precipitate with CH_2Cl_2 and evaporate the filtrate at 15 mm Hg, 30° C to get LSD.

LSD Via SO_3 METHOD 1

This and the following method are expanded versions of Garbrecht's method. My own prejudice is that it makes much more sense to use one of the other, simpler methods since the unreacted lysergic acid can be recycled, and the initial yield is consequently of little import. However, the details as presented here derive from the practical experience of underground chemists and contain many points of interest for any technique.

Notes on Processes

1. Chemicals to process one kilogram of ergotamine tartrate:

```
Alumina, Activity II, 100-200 mesh ............ 8 Lbs.
Benzene, Reagent ......................... 20 Liters
Charcoal, activated powder, Norit A ........ 100 Grams
Congo red papers ............................. 1 Vial
Dichloromethane (methylene chloride), Purified .. 60 Liters
Diethylamine, Reagent.......... 500 Grams or 725 ml.
Dimethylformamide, Reagent ................ 12 Liters
Ergotamine tartrate ..................... 1 Kilogram
Ether, Absolute ............................. 5 Lbs.
Ethyl alcohol, Anhydrous, Denatured .......... 10Liters
Lithium hydroxide hydrate, Reagent ......... 200 Grams
Methanol, Reagent........................... 24 Pints
Molecular sieve, Linde 4 A ................... 1 Lb.
Petroleum ether, B.P. 40°-70° C ............. 2.5 Liters
```

Phenolpthalein, White . . . . 1 Gram
Phosphorus pentoxide, Reagent . . . . 100 Grams
Potassium hydroxide, Reagent . . . . 2 Lbs.
Sodium Chloride, Reagent . . . . 5 Lbs.
Sulfuric acid, Fuming 33% . . . . 3 Lbs.
Sulfuric acid, Reagent . . . . 4 Lbs.
Tartaric acid, natural, powder, Reagent . . . . . . 200 Grams

2. Cylinder gasses:

Ammonia, anhydrous . . . . 1 Lb.
Nitrogen, Dry . . . . 1 small welding cylinder

3. Notes on preparing reagents:

Ammoniacal ethanol is prepared by chilling ten liters of anhydrous denatured ethyl alcohol as commercially purchased in a freezer to well below 0° C. Next, 600 to 750 ml of liquid ammonia is drawn from a pressure cylinder into a 1000 ml graduate in a well ventilated area. The contents of the graduate are carefully poured into the chilled alcohol. The solution is then stirred to mix and warmed to room temperature. The solution should be at least two molar as determined by titration against standard acid solution to a methyl red endpoint. If titration is to be attempted, a little methyl red should be added to the chemical list.

Dilute sulfuric acid is prepared by pouring 750 ml of sulfuric acid into about 11.25 liters of ice cold water in a large acid resistant container such as a polyethylene jug or 5 gallon gasoline container, etc. Anhydrous dimethylformamide for making up sulfur trioxide-dimethylformamide reagent is prepared by shaking three liters of reagent dimethylformamide with 100 to 200 grams of Linde molecular sieve, and allowing the mixture to stand overnight with occasional shaking. Next, the dimethylformamide is decanted off and poured into a five liter boiling flask. The flask is fitted with a helices packed fractionating column and distilled at 25 millimeters pressure. 800 ml is distilled off as a forerun and discarded or re-dried over molecular sieve. The next fraction of about two liters is collected and kept so as to protect from atmospheric moisture (drying tube, etc.). A little dimethylformamide is left in the boiling flask. Dimethylformamide when prepared in this manner is excellent for preparing the sulfur trioxide reagent. Commercially available spectro-quality or pesticide quality dimethylformamide may also be used if the water content is specified to be less than .05%, but the reagent obtained from these products has always appeared darker in color than that made by the above method. Commercial reagent quality dimethylformamide is suitable for the main reaction.

4. Preparation of the sulfur trioxide-dimethylformamide reagent (SO_3·DMF).

Sulfur trioxide is distilled from fuming sulfuric acid. About 200 grams are necessary and can be obtained from varying amounts of fuming sulfuric acid depending upon the concentration of sulfur trioxide in the commercially available product. Occasionally it is possible to purchase fuming sulfuric acid which contains as much as 70% sulfur trioxide (SO_3). Fuming sulfuric acid containing 30-33% is usually easily available and 200 grams of SO_3 can be obtained by distilling about one kilogram of it. The distillation is done at atmospheric pressure in a simple distillation apparatus which utilizes a large bore condenser of the West or similar type. The receiving flask should be connected to the condenser in a closed fashion as in a vacuum distillation and vented to the atmosphere through a drying tube or similar device. Corks, rubber stoppers and any kind of joint grease cannot be used in contact with sulfur trioxide or fuming sulfuric acid as they will char. Tapered glass fittings should be used throughout without grease. All fittings should be cleaned in benzene to remove grease before distilling. Sulfur trioxide fumes are *very* irritating and good ventilation is a must. Glass stoppers and small empty flasks with tapered fittings should be used to close openings in the distilling apparatus when changing receiving or boiling flasks. This will help considerably to keep fumes out of the atmosphere.

Experimental procedure:

200 grams of sulfur trioxide are distilled from the appropriate amount of fuming sulfuric acid. The boiling range should be 5° or less. The 200 grams are collected in a small round bottom flask previously weighed with a glass stopper in place. Next, a small unweighed portion of phosphorus pentoxide is added to the sulfur trioxide in the flask (1 teaspoonful). The flask is swirled and then placed on the distilling apparatus as the boiling flask. A forerun of 5 ml is distilled into a small receiving flask. This is returned to the boiling flask and the apparatus is fitted with a receiving flask (3 liter boiling flask) already containing 1500 ml anhydrous dimethylformamide and a magnetic stirring bar teflon coated. The flask is surrounded by an ice water bath in a nonmagnetic container and a magnetic stirring motor is placed underneath to rotate the stirring bar in the flask. The remaining sulfur trioxide is distilled into the

receiving flask containing the dimethylformamide. The receiving flask should be absolutely dry before filling it with the dimethylformamide. The sulfur trioxide should distill off over a 2° boiling range the second time. A hot water bath or an oil bath is convenient when heating the boiling flask and prevents overheating. When the distillation is complete, the receiving flask is removed carefully and stoppered. The flask is warmed and swirled to dissolve any crystalline material and then cooled to around 5° C. and left several hours. If any material precipitates, a little anhydrous dimethylformamide is added to dissolve the residue. The flask is swirled and the contents decanted into a storage bottle. The temperature in the storage bottle is recorded and a ten milliliter aliquot withdrawn with a pipette. The aliquot is run into a 250 ml Erlenmeyer flask and diluted with 10 ml water to decompose the complex. The mixture is then titrated to a phenolpthalein endpoint with a standard base solution. A convenient standard base solution can be made by dissolving 1 mole of lithium hydroxide hydrate (41.96 gr) in distilled water to make one liter in a volumetric flask. Three consecutive titrations should be done and an average taken. One mole of sulfur trioxide reacts with two moles of lithium hydroxide. The reagent bottle should be labeled as to sulfur trioxide concentration (about 1.5 molar) and the temperature at which the concentration was determined since the reagent has a rather high coefficient of expansion.

If at any time during the distillation of the sulfur trioxide, crystals of solid sulfur trioxide form in the condenser or receiving flask, they may be melted by careful local heating with a propane torch flame or by running hot water through the condenser jacket. The water in the condenser should be above 23° C during distillation of sulfur trioxide to prevent crystallization of sulfur trioxide polymers.

5. Notes on changing the scale of reactions:

The ergotamine to lysergic acid reaction may be scaled up or down by multiplying the quantities involved by a proportionality constant: all quantities should be multiplied by the same constant. It has been found that the quantities of water and potassium hydroxide used in the hydrolysis of ergotamine are not particularly critical and their relative concentrations may be varied somewhat to meet other considerations. As a rule, ergotamine should be hydrolysed with about a 1.5 to 2.5 molar potassium hydroxide solution.

The lysergic acid to lysergic acid amide reaction has been designed to utilize minimal quantities of solvents in order to squeeze as much material as possible into ordinary laboratory glassware.

Some workers have suggested that the quantity of dichloromethane (or chloroform) can be further reduced and still effectively extract the amide, but this may prove difficult, especially if emulsions are encountered. When scaling down the reaction, if desired, the quantities of methanol, dimethylformamide, saline solution, and dichloromethane may be in greater quantity than calculated by direct proportion to the other reagents. The proportional relationship between lysergic acid, lithium hydroxide and sulfur trioxide must be strictly adhered to. The molar proportions are: lysergic acid, 1 mole; lithium hydroxide, 1 mole; sulfur trioxide, 2 moles. Diethylamine should be added in at least five molar equivalents. 6.5 equivalents are used in the example given. In general, two thirds of the dimethylformamide should be distilled off from the lithiumlysergate solution. It is convenient to do the reaction in a small quantity of dimethylformamide if doing large quantities of lysergic acid since the product is contained in smaller volume and extraction may be done with less solvent.

6. Notes on purification of amide

The chromatography detailed in the example has been used and works fairly well, however, the removal of all colored impurities is not achieved and there is room for improvement. It is suggested that further experimentation be done to improve the process. An ultraviolet light is indispensable when doing experimentation with these compounds. The lysergic acid amide displays a blue fluorescence. Benzene has been used successfully as an eluant on activity 4 alumina, but the results were no better than the example given. Chloroform and chloroform-benzene mixtures also have been used on varying grades of alumina but no useful data is available.

7. Notes on crystallization of tartrate

Methanol and methanol-ether mixtures have both been used to crystallize the amide tartrate. Crystallization proceeds more readily if ether is present. Usually, the quantity of solvent from which the tartrate is crystallized should be around 4.0 to 6.0 times the weight of the free base amide (in grams) expressed in milliliters. For example, if the free base lysergamide weighs 20 grams, then the crystallization should be done in 80 to 120 milliliters of solvent. The purer the free base amide, the less solvent that may be effectively used and the higher the yield. The solvent is reagent grade methanol containing 10.0% to 25% ethyl ether. It is usually preferable to dissolve the amide in the methanol and then to add the ether. Ether should not be added after the tartaric acid is added since it precipitates the impurities at the site of addition. Crystallization occurs more slowly with impure preparations. Considerable time should be allowed in the cold before filtering. Overnight is excellent.

Lysergic Acid Monohydrate

350 grams of potassium hydroxide are dissolved in 3500 ml of water in a five liter three-neck boiling flask equipped with a reflux condenser and a small tube introducing a slow stream of nitrogen gas. The mixture is then heated to about 80° C, when 500 grams of ergotamine tartrate is added to the flask. The temperature is maintained at 80° for about 2½ hours while continuously bubbling nitrogen gently through the mixture.

The reaction mixture is next poured into a ten liter polyethylene bucket and diluted by addition of ice to an approximate volume of six liters. The bucket is then placed in an ice water bath and the mixture cooled below 10° C, after which the mixture is slowly neutralized by the careful addition of cold *dilute* sulfuric acid (15 parts water: 1 part acid) to a congo red endpoint (pH 4.0 - 4.4). Lysergic acid and considerable potassium sulfate precipitates at this time. The bucket is allowed to stand in the ice water bath several hours when the precipitate is filtered off on a large buchner funnel (15 cm or larger) and sucked as dry as possible on the funnel. The slightly moist filter cake is broken up and placed in a four liter beaker containing 2½ liters of two molar ammoniacal ethanol (about 300 ml liquid ammonia poured into 5.0 liters of *chilled anhydrous* denatured ethanol). The contents of the beaker are stirred for one hour and then filtered. The filtrate is kept and the filter cake is broken up and extracted in the previous manner with a second portion of two liters ammoniacal ethanol. This extract is filtered using 500 ml of ammoniacal ethanol to wash down the filter cake and the combined filtered extracts are taken to total dryness in a rotary vacuum evaporator over a boiling water bath. The tan colored residue is easily scraped from the sides of the evaporator flask by means of two bent wire rods, one bent less than 90° and one bent greater than 90°. The residue is scraped into a large mortar as well as possible with the bent rods and then 225 ml of methanol mixed with 75 ml of water is used in divided portions to wash the remaining residue in the flask into the mortar. The last portion of the methanol-water mixture (about 100 ml) is left in the evaporator flask to be used to wash the mortar clean later. The slurry in the mortar is ground with a pestle to an even consistency free of lumps when it is then poured and scraped into a large Buchner funnel (15 cm or larger) and filtered. The remaining portion of the methanol-water mixture is used to wash down the mortar and the filter cake. Next the filter cake is washed with at least 250 ml of water and sucked dry for an hour. The filter cake is broken up and dried to a constant weight under high vacuum at 90° C in a dessicator.

Yield: 125 to 150 grams of lysergic acid monohydrate, MW.286.35. A slightly off-white powder.

N,N-Diethyllysergamide (LSD)

143.20 grams of lysergic acid monohydrate (0.5 mole) and 21.0 grams of lithium hydroxide hydrate (0.5 mole) are dissolved in 2500 ml of methanol with stirring and warming in a four liter beaker. When the lysergic acid is completely dissolved, the contents of the beaker are admitted to a rotary vacuum evaporator and taken to total dryness over a boiling water bath. A little methanol is used to rinse any lysergate residue that may remain in the beaker into the evaporator. The crumbly, tan colored dry residue is dissolved and rinsed into a five liter boiling flask with three liters of anhydrous dimethylformamide. Considerable care should be exercised when transferring solutions from one vessel to another to avoid loss of lysergate since small deviations from the calculated quantities of reagents result in considerable reduction in overall yield. The five liter flask is next fitted with a 600 mm helices packed fractionating column and about 2050 ml of dimethylformamide is carefully distilled off at 10.0 millimeters pressure to remove water from the lysergate solution. The boiling flask containing lithium lysergate in the remaining dimethylformamide is tightly stoppered and chilled in an ice water bath to below 5° C. 1.0 mole of sulfur trioxide is now added to the flask by addition of the appropriate amount of sulfur trioxide-dimethylformamide reagent (previously prepared by double distilling sulfur trioxide from fuming sulfuric acid and slowly adding it to anhydrous dimethylformamide to make a solution of approximately 1.5 molar strength as determined by titration against standard base solution). Cooling and swirling are continued for 15 minutes when 335 ml of diethylamine is added. Cooling and swirling are continued 15 minutes longer when the reaction mixture is poured into 3800 ml of a 20% saline (sodium chloride) solution to break the reaction complex. The reaction mixture is now extracted with 10.0 to 12.0 liters of methylene chloride (dichloromethane) or chloroform in divided portions in a separatory funnel. A scheme for division of the extraction solvent is as follows:

Extract	Quantity	Total Solvent Used
First	2000 ml	2000 ml
Second	1800	3800
third	1500	5300
Fourth	1200	6500
Fifth	1000	7500
Sixth	1000	8500
Seventh	800	9300
Eighth	700	10,000
etc.	etc.	etc.

Continue until clear lower layer appears.

The extracts from the reaction are combined and shaken up with a little anhydrous magnesium sulfate (120 grams) and filtered. The filtrate is evaporated to dryness in the rotary vacuum evaporator, care being taken not to heat the extracts or the residual syrup above 55° C. A good mechanical vacuum pump and effective cold traps in the line are necessary to remove the residual dimethylformamide from the residue. A brown to black bubbly residue should remain when evaporation is complete. This residue contains the amide product and considerable impurities. A general method of purifying the amide follows.

Method A

The material to be purified (the above residue or other material containing N,N-diethyl lysergamide) is taken up in 1200 ml of methylene chloride containing 20% benzene and applied to a chromatographic column containing two pounds of basic alumina 100-200 mesh Brockmann activity *two* or *three*. The column is eluted with nine liters of methylene chloride containing 20% ben-zene. At this point, the column when viewed with visible light should display three distinct color bands. The uppermost band will be a dark brown or greyish color, the next band will be a reddish brown color, and the lowest band will be a light brown or tan color. The eluant should be amber colored. The column may now be eluted with about one liter methylene chloride containing 0.5% methanol. This will bring the reddish band nearly to the bottom of the column. At no time should any portion of the reddish band be eluted from the column. If any of the reddish band reaches the bottom of the column, elution should be stopped. Next, the total eluant is shaken up with 30 grams of activated charcoal (Norit A) mixed with 30 grams of alumina and filtered. The filter is washed with 600 ml of methylene chloride and the total filtrate taken to dryness on the rotary vacuum evaporator. Care is taken not to heat the residue or solution above 55° C. The residue is taken up in one liter of benzene and immediately taken to near dryness when another liter of benzene is added to dissolve the residue and the solution is again taken to near dryness. This procedure is repeated until four to five liters of benzene have been added and evaporated. The residue is finally taken to *complete* dryness at about 45° C. If sufficient benzene has been added and evaporated, a light tan bubbly, crystalline material will fill the interior of the evaporator flask. It is important that this residue be completely dry before proceeding. The evaporation of benzene from the residue aids removal of solvents and other volatile materials (as aseotropes) which promotes formation of the bubbly crystalline structure in the residue. 700 ml of petroleum ether is next added to the evaporator flask which is then removed from the evaporator and tightly stoppered. The flask is shaken vigorously to loosen the residue from the sides of the flask. Usually all the material comes loose from the flask and forms a slurry in the petroleum ether. If necessary, a bent wire rod may be used to scrape material from the flask. The slurry is now decanted into a buchner funnel and filtered. The filtrate is used to further wash material from the flask into the filter funnel. The filter cake is sucked as dry as possible and then dried to a constant weight under high vacuum at 45° C in a dessicator.

Yield: approximately 130 grams N,N-diethyllysergamide MW 323.42.

The material remaining on the column may be removed with methanol, evaporated in a vacuum and recycled through the isomerization and subsequent procedures by itself or combined with fresh material. Also, all leftover solutions and residues may be neutralized with sodium bicarbonate, evaporated in vacuo, extracted with ammoniacal chloroform, the extract evaporated to dryness, and the residue re-used.

N,N-Diethyllysergamide Tartrate

130 grams of N,N-diethyllysergamide is dissolved in 400 ml methanol and filtered. The filter is washed with 30 ml methanol and the filtrate and washing is poured into a one liter beaker. 30 ml more methanol is used to further wash the filter and filter flask and the wash is also poured into the beaker. 130 ml of diethyl ether is now added to the contents of the beaker. The beaker is gently warmed on a hot plate and 32.0 grams of tartaric acid are then added with constant stirring and warming until they are completely dissolved.

The beaker is then allowed to cool. Crystallization of the tartrate usually begins as soon as the tartaric acid dissolves completely. The beaker and contents are refrigerated for at least four hours. Occasional stirring of the crystallizing solution will produce smaller crystals, whereas if the solution is left unstirred during the crystallization, larger crystals will grow. Either is satisfactory. After the beaker has been allowed to stand in the cold four hours or more, the contents are filtered off on a 110 mm buchner funnel with suction. The crystals are washed on the funnel with first 200 ml of a two part methanol:one part ether mixture, and then with 250 ml of a two part ether:one part methanol mixture. Next the crystals are washed with 600 ml of ether and sucked dry. The filter cake is broken up and allowed to air dry in a warm, dark place.

First crop yield: Approximately 80 grams pale yellow to white needles.

The mother liquors and the two washes containing methanol are collected and combined. A one normal solution of potassium hydroxide in methanol is added in approximately equal volume to the combined washes and mother liquors. The solution is then filtered and the filter washed with a few ml of methanol. The filtrates are allowed to stand at room temperature for two to three hours to re-equilibrate the iso-lysergic acid amides from the mother liquors. About 500 ml of water is then added and the mixture extracted with 2.5 liters of methylene chloride in divided portions in a separatory funnel. The combined extracts are shaken with 25 grams of anhydrous magnesium sulfate and filtered. The filtrate is taken to dryness on the rotary vacuum evaporator, care taken not to heat above 55° C. The material is purified in the same manner as that from the original reaction mixture using approximately one fourth the quantities of solvents and alumina as for the original.

Second crop yield: Approximately 20 grams white needles.

The mother liquors may again be worked up as before, or alternatively, they may be saved and included in subsequent batches.

Third crop yield: Approximately 5 grams white needles.

Total yield: Approximately 105 grams N,N-diethyllysergamide tartrate MW = 430.51 (includes one mole methanol per mole of amide).

Method B

The residue from the previous step is taken up in two liters of chloroform and filtered with suction through a column 50 millimeters in diameter packed with 400 grams of basic alumina, Brockman activity 1. The filtrate is then refiltered through the same column in the same manner four or five times until the filtrate appears light amber and further repetition of this process fails to remove significant color from the filtrate. The column is now eluted by adding several liters of fresh chloroform to the top and sucking it through into the previous filtrate. Sufficient chloroform should be added to remove all blue fluorescent material from the column but not greenish or yellow (use a blacklight in a darkened room). A band of greenish yellow material should remain in the upper 2/3 of the column when viewed in ultraviolet light (blacklight). The total filtrate is taken to dryness in vacuo in a three liter round bottom flask on a rotary evaporator over a 60^0 hot water bath. The residue is taken up in 500 ml of benzene and again taken to dryness in the same manner. 500 ml of benzene is again added and taken to dryness. The flask is left on the evaporator under full vacuum for a considerable length of time after the residue appears dry to remove any traces of dimethylformamide that may still remain. A bubbly, crystalline residue should fill the interior of the flask at the end of this step. If any tarry, gummy appearing material appears to remain on the sides of the flask, repeat the addition of 500 ml of benzene and evaporate to dryness again to get a glassy, crystalline appearance. When the material in the flask is totally dry, remove the flask from the evaporator and add sufficient petroleum ether (a commercial mixture of hexanes is excellent for this purpose) to the flask to be able to swirl the crystalline material around and loosen it from the sides of the flask. Filter this slurry on the buchner funnel with a fritted glass disk and use the filtrate to further wash the remaining material from the evaporator flask into the buchner funnel. Suck the material dry on the funnel and then place in a vacuum dessicator and dry to a constant weight. Record the dry weight of this material N. Calculate the weight of one equivalent of tartaric acid as follows:

$$\text{Weight of tartaric acid} = .232N$$

Add methanol to a small beaker in a quantity equal to four times N in milliliters. Dissolve the dried material of weight N in this. Dissolve one equivalent of tartaric acid in the same solution, warming the solution gently and stirring. Slowly with stirring, add ether to the solution in the quantity of no greater than .5 N ml. Addition of ether causes a precipitate which dissolves quickly. Ether should be added dropwise with stirring between drops to dissolve any precipitate before addition of the next drop. Crystallization of LSD tartrate should begin shortly after or during addition of the ether. This precipitate does not dissolve and should not be confused with the precipitate caused by the addition of ether. The mixture should be stirred until the solution becomes thickened by formation of crystals. Once crystallization of LSD tartrate is begun it is unnecessary to continue addition of ether. The beaker should be refrigerated several hours and the contents then filtered on a buchner funnel with a fritted glass disk. The crystals are sucked dry and washed with 2.0 N milliters of methanol previously chilled below -5° and then with 4 N milliliters of a 1:1 mixture of cold ether and methanol. The crystals are sucked completely dry, washed with 8 N ml of ether, sucked dry, and placed in a vacuum dessicator to remove last traces of solvent.

The total filtrates from the crystals (mother liquors plus washings) are made basic by addition of 2 molar ammoniacal ethanol in approximately equal volume and allowed to stand several days at room temperature when the mixture is filtered and taken to dryness and treated in the same manner as the residue from step 2 for a second crop of crystals.

LSD Via SO₃ Method 2
Lysergic Acid

Ergotamine tartrate (10g) is added to a stirred de-aerated (nitrogen stream) solution of 38 g potassium hydroxide in 100 ml of methanol and 200 ml of water. The solution turns pink to red. The solution is heated to reflux and the methanol is slowly removed using a partial takeoff. Methanol is allowed to distill until the pot temperature reaches 90-95° C. The mixture is then maintained at total reflux until the evolution of ammonia ceases (hold pH paper in outlet of reflux condenser to test for ammonia). Nitrogen should be bubbled through the mixture to entrain the ammonia.

The hot dark solution is then allowed to cool somewhat and then cautiously acidified with a mixture of 60 ml acetic acid and 60 ml water. The resulting hot solution is quickly treated with Norite "A" decolorizing carbon and filtered hot.

The clear purple-hued filtrate is allowed to cool to room temperature (crystallization begins) and then in an ice bath or refrigerator. The crystalline precipitate of lysergic acid (grey to purplish white) is collected, washed with a small amount of cold water (5 ml), followed by cold methanol (5 ml) and ether. Yield 3.2

to 3.8 g.

Digestion of the crude acid with about 50 ml of methanol (to remove some of the colored impurities) gave after cooling to 0-10° C and filtering an almost quantitative recovery of lighter colored acid. This material is suitable for conversion into LSD.

Lysergic Acid Diethylamide

1. Sulfur Trioxide-Dimethylformamide Complex

Into a carefully dried two liter three necked round bottomed flask fitted with a mechanical stirrer, thermometer and a pressure equalizing dropping funnel protected from the atmosphere with a $CaCl_2$ drying tube, was placed approximately one (1) liter of freshly distilled dimethylformamide (DMF) (a one to three degree fraction BP ca. 62-63° C/20 mm). The DMF was cooled to 0-5° C by means of an external ice-salt-water bath. Sulfur trioxide (Sulfan B) (ca. 100 g) was then placed in the dropping funnel and added dropwise over a period of 30 to 40 minutes to the stirred DMF. The temperature is carefully maintained between 0 and 5° C throughout the course of the addition. Stirring is continued thereafter until all of the crystalline material is brought into solution.

The resulting reagent solution is then transferred into a suitable reservoir fitted with an automatic burette (protected from the atmosphere with a Drierite tube) and refrigerated. If kept dry, the reagent will be good for a month or two even though it will turn yellow and then orange in color.

The molarity of the reagent is then determined by titration against standard base. An aliquot (1 or 5 ml) is first diluted with water (20 or 100 ml) to convert the sulfur trioxide-DMF complex into sulfuric acid. The resulting solution is titrated to phenolphtalein end-point with standard 0.1 or 0.01 N aqueous alkali (NaOH or KOH) to determine the *molarity* (½ of the Normality). It should be in the range of 0.9 to 1.2 depending on the amounts of SO_3 and DMF used.

2. Lysergic Acid Diethylamide (LSD)

For best results all lysergic acid and LSD solutions should be protected from direct light (yellow light is non-damaging) and the working temperatures should never exceed 25° C.

Lysergic acid monohydrate (7.15 g, 25.0 mmol on a 100% basis) and lithium hydroxide monohydrate (1.06 g, 25.0 mmol) were added to 200 ml of anhydrous methanol and stirred until complete solution occurs. Use magnetic stirrer and keep solution

131

under dry nitrogen in the dark. The solvent methanol is then removed by evaporation under reduced pressure to leave a frothy glass-like residue of lithium lysergate.

A solution of the calculated amount of tartaric acid is prepared in methanol (*ca.* 8 ml/g). Approximately ½ of the methanol to be used and 20% of the tartaric acid solution is added to the flask containing the LSD base. The flask is swirled and/or shaken until the solid material has dissolved (5-10 minutes) and the solution is then transferred into an Erlenmeyer flask. The balance of the methanol, in two portions, is used to complete the transfer. At this point the rest of the tartaric acid solution is added. It may be helpful to titrate the solution to an end-point pH of 5.3, since adding excess tartaric acid solution inhibits crystallization somewhat. However this is optional. If seed crystals are available, they should be added at this point. Crystallization should begin within a ½ hour. The flask should then be refrigerated for 12-24 hours at 5-10⁰ C and then for another 12 hours at -10 to -20° C. For 5 g of LSD base 1 g of tartaric acid in 7-8 ml methanol and an additional 17-18 ml of methanol are used.

The crystalline mass of needles is broken up and the cold solution filtered (suction). The filter cake is sucked dry and then washed with anhydrous ether. If necessary the product may be recrystallized from methanol using 5 ml for each gram. The snow white product melts at 198-200° C.

3. Recrystallization Procedure

The crude tartrate (10 g) is placed in a 125 ml Erlenmeyer flask and boiling methanol (50 ml) is added and the mixture stirred and heated for a minute or two (*no longer*) until solution is complete. The hot solution is quickly filtered through a previously warmed buchner funnel and the filtrate cooled *immediately* by swirling in a cold water bath until the temperature drops to 25° C. Crystallization should be well on the way by this time. The mixture is further cooled to 5 to 10° C and then to -10 to -20° C as previously described, to complete the crystallization. Recovery is between 50 and 70%.

4. Additional Crops of Crystals

The mother liquors from initial crystallizations and from re-crystallizations of LSD can be concentrated by evaporation under reduced pressure to produce additional crops of crystals. The second and third crops of crystals are usually dirty enough to require re-crystallization. After three crops, the mother liquors usually become very syrup-like. They then contain mostly iso-LSD (as the tartrate salt).

The iso-LSD salt can be converted back into the base by the addition of methanolic KOH or potassium methoxide to the mother liquor. The resulting mixture should be added to a separatory funnel containing salt solution and ethylene dichloride. The LSD base is extracted into the ethylene dichloride layer (the lower layer). The lower layer is removed and fresh ethylene dichloride used to extract the last traces of LSD base from the salt water-base mixture. The ethylene dichloride extracts are combined, dried with $MgSO_4$, decolorized and filtered through diatomaceous earth as earlier. The resulting ethylene dichloride solution may be combined with the chloroform solutions of iso-LSD which eluted from the chromatographic column. The combined solution may be evaporated to dryness under reduced pressure.

5. Isomerization

The dry iso-LSD base can then be dissolved in methanol and potassium methoxide added. The resulting mixture is stirred for about 30 minutes. During this time isomerization takes place; about 70% of the iso-LSD is converted into the desired "normal" form of LSD.

The methanolic solution is poured into a separatory funnel containing salt water solution and ethylene dichloride. The salt water layer is repeatedly extracted with ethylene dichloride to separate the LSD base from the water-base mixture. The ethylene dichloride extracts are combined, dried with $MgSO_4$, decolorized and filtered. The ethylene dichloride solution is then evaporated to dryness under reduced pressure.

The resulting dry LSD base is chromatographed on basic alumina (activity grade 1) as previously described. The blue band is collected as before, evaporated and converted into the tartrate salt. The iso-LSD band may be collected and saved for further re-cycling.

NOTE: If you only have mother liquors to isomerize, the second mixing with potassium methoxide is unnecessary. Simply prolong the initial mixing to about ½ hour.

Total Synthesis of Lysergic Acid

Of the many attempts at the total synthesis of lysergic acid from simple starting materials, only two have been successful (JACS 78,3087(1956), which is *very* complicated, and CA 74,3762 (1970), which follows). However, it is very likely that some of the intermediates in each attempt are psychedelic. In fact, this is one of the most promising and least investigated areas of psychedelic chemistry. Following are some references to the synthesis of intermediates: CCCC 33,1576(1968); HCA 33,67,375,1796, 2254,2257(1950), 34,382(1951), 35,1249,2095(1952), 36,839 1125,1137(1953), 37,1826(1954), 38,463,468(1955), 44,1531 (1961); JCS 3399,3403(1954); BSC 861(1965); JOC 10,76 (1945), 26,4441(1961), 29,843(1964); Chem. and Ind. 1151 (1953); CJC 41,2585(1963); CPB 12,1405,1493(1964), 13,420 (1965), 14,1227(1966); JACS 61,2891(1939), 67,76(1945), 71,761(1949), 73,2402(1951), 78,3087(1956), 79,102(1957), 82,1200(1960), 88,3941(1966); BER 86,25,404(1953), 87,882 (1954), 88,370,550(1955), 89,270,2783(1956), 90,1980,1984 (1957), 93,2024,2029(1960), 96,1618(1963), 100,2427(1967), 101,2605(1968); JMC 8,200(1965); C.R. Acad. Sci. Paris 264 (C), 118(1967), 265 (C), 110(1967); BSC 1071(1968); CJC 41,2585(1963); CPB 12,1405,1493(1964), 13,420(1965), 14, 1227(1966); JCS (P.T.I.) 1121(1972), 760(1973), 438(1973); CA 78,71830,72200(1973).

Various analogs containing part of the LSD structure have been synthesized, but few have any activity. See JPS 60,809(1971) for a review of these compounds.

For other LSD analogs see JMC 16,804,1015(1973); BSC 2046(1973); CA 48,4489(1954); JPS 62,1881(1973).

Some other useful references on LSD chemistry: U.S. Patents 3,856,821 and 3,856,822; Swiss Patent 517,680(1970); Belgian Patent 738,926; French Patent 1,368,420 and addition 91,948 (1968).

Total Synthesis of LSD German Patent 1,965,896 (1 Oct 1970). German Patent 1,947,063 is the same as 1,965,896.

For synthesis of 5-Br-isatin from isatin see CA 33,2516(1939) or BER 40,2492(1907).

For the use of cycloaliphatic or aromatic esters in place of methyl-6-methylnicotinate or of isatin or 4 or 5 chloroisatin or 4-bromoisatin in place of 5-bromoisatin see French Patent Addition #2,052,237 (14 May 1971).

Total yield of LSD from starting materials is probably about 1%.

Mix 32.8 g (0.217M) methyl-6-methylnicotinate (other alkyl groups can replace either methyl group) with 45.2 g (0.2M) 5-bromoisatin (apparently 4-Br or 4 or 5 Cl isatin will also work) in a 250 ml flask at 100° in an oil bath and raise the temperature of the bath to 180° over 15 minutes. Lower temperature to 170° and let react for 70 minutes. Cool and then grind the solid as fine as possible in a mortar. Recrystallize from 150 ml dimethylformamide and wash with ether to get 40 g (57%) methyl- α (5-bromo-3-isatylidene)-6-methylnicotinate (I). Suspend 10 g (I) in 250 ml glacial acetic acid and heat to boiling. Add in small portions over 30 minutes excess powdered zinc. Reflux 1 hour, filter and evaporate in vacuum and recrystallize the residue from dioxane to get 9.7 g (95%) methyl- α -(5-bromo-2-oxindol-3-yl)-6-methylnicotinate (II). To a suspension of 18 g dry $NaBH_4$ in 300 ml dry tetrahydrofuran add with stirring at 0° over 30 minutes about 75 g BF_3 etherate. Stir 3 hours at 0°, add 18 g (II) and heat exactly 20 minutes at precisely 22-24°. Add carefully 150 ml concentrated HCl while cooling in an ice bath. Add 200 ml water and stir 12 hours. Basify, extract the product with ethyl acetate and dry, evaporate in vacuum to get 11 g of residue which recrystallizes from methanol to give methyl- α -(2,3-dihydro-5-bromo-3-indolyl)-6-methylnictotinate (III).

The following step may be unnecessary but it gives stability to (III). The acetyl group can be split off at the end of synthesis, but this is unnecessary since the 1-acetyl-LSD is as active as LSD.

Treat 12 g (III) at room temperature for 24 hours with acetic anhydride then hydrolyze and extract to get 11.5 g residue which is ground in petroleum ether and recrystallized from cyclohexane (can chromatograph on alumina and elute with petroleum ether to wash out an oil, then with benzene containing 5% ethyl acetate to elute the produce) to give methyl- α (1-acetyl-5-bromo-2,3-dihydro-3-ind-olyl)-6-methylnicotinate (IV). Heat 5 g (IV), 12.5 ml acetone, 12.5 ml methanol and 1.8 ml methyl iodide for 18 hours in a Carius tube at 70-80°. Cool, filter, wash with acetone and recrystallize from methanol to get methyl- α (1-acetyl-2,3-dihydro-5-bromo-3-indolyl)-1,6-dimethylnicotinate iodide (V). To 9.4 g (V) in 250 ml water and 250 ml methanol at 35° add over 5 minutes 2.9 g KBH_4 and stir 10 minutes. Add 2.9 g more KBH_4 and stir 30 minutes. Evaporate in vacuum and extract the residue with methylene chloride to get 6.2 g oily mixture containing about 2 g of the d isomer (can separate by chromatography if desired) of methyl- α (1-acetyl-2,3-dihydro-5-bromo-3-indolyl)-6-methyl-1,2,5,6-tetrahydro-nicotinate (VI).

To a suspension of finely powdered $NaNH_2$ (6.1 g) in 2 liters dry ammonia, add with stirring 8 g (VI) in 50 ml dry tetrahydrofuran. Stir 1 hour, add NH_4Cl and evaporate the ammonia as fast as possible in a nitrogen stream. Extract at pH 8 with methylene chloride to get 6 g (can chromatograph on 300 g silica gel and 250 g Celite and elute with 98% benzene-2% absolute ethanol and evaporate in vacuum) of methyl-1-acetyl-2,3-dihydro-lysergate (VII). (VII) can be converted to 2,3-dihydro-LSD (not to be confused with 9,10-dihydro-LSD, which is inactive), which is about ten times less active than LSD. (VII) can be converted to lysergic acid prior to conversion to LSD, which will triple the yield in terms of LSD activity (considering 30% yield). The process (which follows) is somewhat complicated and an easier dehydrogenation process may work. 2,3-dihydro-LSD can be converted directly to 12-hydroxy-LSD, which has about the same activity as LSD and this process is also given below.

Lysergic Acid from 2,3-Dihydrolysergic Acid JACS 78,3114 (1956)

Dissolve 4 g (VII) in 78 ml 1.5% KOH and reflux five minutes (under N_2 if possible). Add 8.5 g Na arsenate hydrate and 16 g Raney-Ni (wet) (deactivated by boiling in xylene suspension — see JOC 13,455(1948)) and reflux twenty hours (under N_2 if possible). Filter, precipitate lysergic acid by taking pH to 5.6 with HCl; filter and wash precipitate with water to get 1 g lysergic acid. Evaporate in vacuum the filtrate to get more product.

12-Hydroxylysergamides from 2,3-dihydrolysergamides HCA 47, 756(1964)

Warm to dissolve 1.5 g 2,3-dihydro-LSD in 5 ml acetone, 40 ml water and 40 ml saturated $NaHCO_3$. Cool to 20° and add all at once with vigorous stirring 2.46 g potassium nitro-sodisulfonate dissolved in 90 ml water and 10 ml saturated $NaHCO_3$. After 1 minute, extract 7 times with ethylacetate, wash the combined extracts with water, dry and carefully remove solvent to get a mixture of 12-OH-LSD, LSD and starting material which can be chromatographed to give about 0.2 g 12-OH-LSD.

The following method of converting (IV) to the diethylamide (which can probably be used in place of (IV) to give the diethylamide of (V), (VI), and (VII)) will probably also work admirably for (VII) or lysergic acid.

Reflux 0.5 g (IV) with 0.5 g KOH in 30 ml methanol for 4 hours. Evaporate in vacuum and add water to the residue. Adjust the pH to between 5 and 6 and filter or centrifuge to get 0.3 g of the free acid. Suspend 1.8 g of the acid in 125 ml chloroform, cool to -5° and add

0.5 g triethylamine, then 0.6 g ethylchloroformate and stir 45 minutes. Add 2 ml diethylamine and stir 3 hours at room temperature to get, after the usual workup, 1 g of the diethylamide (recrystallize from benzene).

Summary of Procedures and Materials
For 1 Kilogram Ergotamine

★ *Starred chemicals are carefully watched.*

★★★★diethylamine 500 g(725 ml)
      methylethylamine 50 g(75 ml)
      ethylisopropylamine 50 g(75 ml)
      diethyl ether 5 lbs
      potassium hydroxide pellets 2 lbs
      methanol 5L
      activated charcoal powder (E.G. norit A) 100 g
      tartaric or maleic acid powder 200 g
      small cylinder N_2 or N_2O or freon
      HCL or sulphuric acid concentrated 500 ml
      chloroform 5 L (optional but desireable)
      ethanol 2 gal (optional but desireable)
      NH_4OH concentrated 500 ml (optional but desireable)

IN ADDITION AT LEAST ONE OF THE FOLLOWING
SETS OF REAGENTS

First choice	N,N-dicyclohexylcarbodiimide	500 g
	N-hydroxybenzotriazol	500 g
	(1-hydroxybenzotriazol)	
	tetrahydrofuran	10 L
	methylene chloride	10 L
Second choice	N,N carbonyldiimidazole	200 g
	dimethylformamide	15 L
Third choice	★★anhydrous hydrazine	4 L
	HCl conc.	1 L
	sodium nitrite	1 kg
Fourth choice	★★phosphorous oxychloride ($POCl_3$)	200 ml
	chloroform	10 L
	methanol	5 L
Fifth choice	acetonitrile or dimethylformamide	15 L
	★★trifluoroacetic anhydride	500 g

COCAINE

The leaves of *Erythryoxylon coca*, from which cocaine is obtained, were used by the Incas at least 1,000 years ago. The annual consumption of coca leaf in South America (mainly Bolivia and Peru) by about twenty million users amounts to about 150,000 pounds of cocaine. They chew their leaves with lime (CaO, etc.) which degrades cocaine to ecgonine — but this compound still relieves hunger and fatigue. Cocaine produces little effect orally, so sniffing is the preferred route of administration. Regarding its supposed aphrodisiac properties, it is interesting that one native legend ascribed the magical origin of the coca leaf to the fascination of a conqueror for the bright eyes of a princess who loved him to death. The plant grows in Australia and Africa, as well as in Western South America, and much cocaine was produced in Java. Other species of the genus seem to produce little or no cocaine.

Coca-cola got its name from the coca leaf extract which it contained (as did a variety of wines) until 1904. Neither tolerance nor physical addiction to cocaine seem to occur, so sniffing it occasionally should be quite safe.

Cocaine has structural and pharmacological similarities to the active constituents of belladonna and jimson weed and can likewise space you out in very undesirable ways, expecially if used frequently. Also, as with every drug, some people are very sensitive to it and can become paranoid, etc. with very little exposure. Cocaine base ("free base") is much more euphorigenic than cocaine and consequently much more damaging. It seems to have the addictive pull of heroin for many users and is probably best avoided.

Coca leaf contains l-ecgonine, cinnamyl cocaine and alpha and beta-truxillyl ecgonine (cocamine), which can be converted to cocaine, but d-ecgonine or pseudoecgonine lead to isomers which are devoid of the strong central stimulating effects of l-cocaine. During the process of isolation from the leaf, l-cocaine is converted to l-ecgonine (tropan-3-beta-of-2-beta-carbonxylic acid), which is easily reconverted to l-cocaine (benzoylecgonine methyl ester).

Cocaine Extraction

Cocaine can be extracted from the leaves with almost any organic solvent. Moisten the dried, powdered leaves with Na carbonate solution and extract with cold benzene or petroleum ether. Extract the organic solution with small amounts of dilute sulfuric acid and basify the extract with Na carbonate (the alkaloids precipitate). Dissolve the precipitate in ether, separate the ether from the aqueous Na carbonate and dry and evaporate in vacuum the ether. Dissolve the residue in methanol and heat with sulfuric acid or methanol-HCl; dilute with water and extract with $CHCl_3$. Concentrate and neutralize the aqueous layer and cool to precipitate methylecgonine sulfate, which is converted to cocaine in one step. The alkaloids can also be extracted directly from the powder with dilute sulfuric acid.

Cocaine from Coca Paste

This process (see Bull. on Narcotics 17,29(1965)) is optional since the paste is usually greater than 70% cocaine.

Dissolve 1 g coca paste in 10 ml 3% sulfuric acid, cool to 0° and add with stirring 8 ml 6% $KMnO_4$ and 10% sulfuric acid, 1 ml at a time over one hour. Let stand ½ hour and add powdered oxalic acid with stirring until the precipitate which has formed dissolves. Extract two times with ether, basify the aqueous solution with NH_4OH and extract four times with 18 ml ether. Dry and evaporate in vacuum the ether to get cocaine. The aqueous solution contains ecgonine, which can be converted to cocaine as shown below.

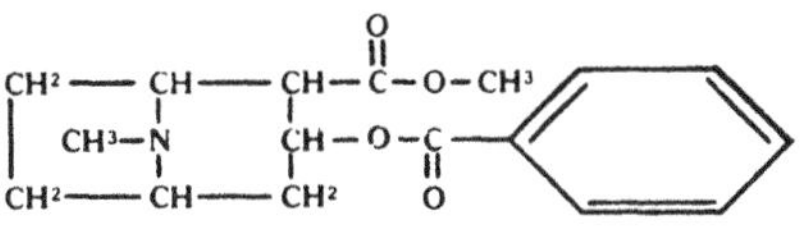

Cocaine

Cocaine Synthesis

The zinc-mercury amalgam used for reduction here and elsewhere can be prepared as follows. Add mossy zinc (powdered will probably do) to 1% aqueous $HgCl_2$, stir awhile, pour off the water and use the Zn-Hg residue for reduction.

Method 1 JGC 30,1138,2070(1960); CA 53,423(1959)

During one hour pass 70 g dry chlorine gas into a stirred solution of 68 g furan in 630 ml methylene chloride and 630 ml methanol at -40°. Protect from moisture ($CaCl_2$ drying tube) and use dry methanol. Alternatively, 160 g Br_2 or Cl_2 in 630 ml methylene chloride (or methanol- in which case add 200 g anhydrous K acetate or pyridine) at -40° is added dropwise to 68 g furan in 630 ml methanol at –40° (keep temperature lower than –30° during the reaction). Stir ½ hour and pass in dry NH_3 (or concentrated NH_4OH) until pH is 8. Filter and wash precipitate with 3X50 ml methylene chloride. If only methanol has been used, after stirring ½ hour pour into 3 L cooled, aqueous, saturated $CaCl_2$ and extract with ether or methylene chloride. Dry and evaporate in vacuum the combined organic solutions to get 90 g 2,5-dimethoxy-2,5-dihydrofuran (I) as a colorless liquid (can distill 40/4) or 71/17). Ethanol can be used in place of methanol, but add 190 ml ether (if reaction temperature rises too high, the methyldiacetyl of maleindialdehyde is produced — this can be reduced to the methyldiacetyl of succinic dialdehyde, which can be used in place of (II) to get (IV)).

Hydrogenate 47.5 g (I) in presence of 5 g Raney-Ni at room temperature and 1 atmosphere H_2 with stirring ($NaBH_4$-Ni reduction described at start should also work). After absorption of 7.2 L H_2 over two to three hours, filter and wash catalyst with 15 ml ethanol and evaporate in vacuum (or distill 77/20 for ethoxy compound, 53/22 for methoxy compound) to get 40 g 2,5-diethoxy (or dimethoxy)-tetrahydrofuran (II).

To a mixture of 360 g 50% KOH and 138 ml methanol, add with stirring at -5° 70.5 g dimethyl ester of acetone dicarboxylic acid (dimethyl-beta-ketoglutarate — see method 3 for preparation) and let temperature rise to about 25° over ½ hour. Let stand ten minutes, cool to 0° and add 65 ml ether. Filter, wash precipitate with 65 ml ethanol and 150 ml ether at 0° to get 75 g (III). To 322 ml 1N HCl at 80°, add 41.1 g (II) and stir twenty minutes; cool to 10°, add 211 ml 1N HCl, 98.2 g (III), 26.4 g Na acetate and 28.2 g methylamine HCl. Stir four hours at room temperature, cool to 10⁰, and saturate with 410 g KOH. Extract four times with methyl-Cl or benzene (75 ml each, fifteen minutes stirring) and evaporate in vacuum to get the methyl ester of tropan-3-one-2-COOH (IV), which precipitates from the oil (can distill 85/0.2). Test for activity. Dissolve 28.3 g (IV) in 170 ml 10% sulfuric acid; cool to -5° and treat with 3.63 kg 1.5% Na-Hg amalgam with vigorous stirring at 0°. See below for easier methods of reducing (IV).

Keep pH about 3.2 by adding 30% sulfuric acid and continue stirring ½ hour or until 3 drops of the mixture fail to give a red color with a 10% solution of $FeCl_3$. Filter and saturate the solution with

235 g KOH or K carbonate below 15°. Extract with 5X250 ml CHCl₃ and dry, evaporate in vacuum to get an oil. (The inactive isomer can be separated at this point (if desired) by letting stand five days at 0° (methyl ester of racemic pseudo-ecgonine precipitates)). Mix the oil plus any precipitate with an equal volume of ether and filter. Add 250 ml dry ether until no more precipitate forms, then filter (test precipitate for activity — if active, this step is unnecessary) and stir with activated carbon ½ hour. Filter and evaporate in vacuum, dissolve the brown liquid in 17 ml methanol, and neutralize with 10% HCl in dry ether. Evaporate the ether until the two layers disappear and let stand two hours at 0° to precipitate the racemic methylecgonine (V). Filter, wash with 1:1 methanol:ether at 0°. Can purify by dissolving in methanol and washing with 1:1 methanol:ether and ether. To prepare the Na amalgam, use an electrolyzer with an Hg cathode, Ni anode, and 40% NaOH solution; current about 29 amps, 7.5 volts.

Heat 9.35 g (V) on water bath ten hours with 18.7 g benzoyl-Cl and pour the liquid formed into 250 ml ether; evaporate in vacuum. The powder formed on rubbing the residue is dissolved in 85 ml ice water and neutralized with 20% NH_4OH. Racemic cocaine (VI) is filtered off, washed with 12 ml ice water and dried over $CaCl_2$. To get the HCl of (VI), dissolve it in seven times its weight of ether containing HCl in ethanol and wash the precipitate with 1:3 methanol:ether, then ether.

Alternatively, mixture of 4 g (V), 36 ml benzene, 1.6 g Na carbonate and 7 ml benzoyl-Cl; stir and heat 96-100° for ten hours. Evaporate in vacuum, cool to 0° and add 40 ml ice water. Acidify with HCl to pH 5 and extract with 3X20 ml ether. Neutralize the aqueous solution with 20% NH_4OH to separate an oil from which cocaine precipitates on standing (from JGC 30,3228(1960)).

For an electrolytic method of producing (I) see ACS 6,531 (1952). For other methods of synthesizing (II) see JACS 72,872 (1950), CA 42,2992(1948). For the reduction of the tropinone (IV) to ecgonine (V), lithium aluminum hydride or $NaBH_4$ give about 50%, the Na-Hg method described above about 40%, and Al triisopropoxide about 25% of the inactive pseudoecgonine. The latter method, which appears to be the best, involves heating at 82° 1½ hours in isopropanol with Al triisopropoxide (Chem. and Ind. 664(1957)). For $NaBH_4$, reflux six hours in methanol. A method said to be superior to that given above for the conversion of methylecgonine to cocaine follows.

Cocaine from Ecgonine BER 89,679(1956)

Dissolve 8.66 g ecgonine in 100 ml methanol and bubble dry HCl gas through for ½ hour. Let stand two hours at room temperature and then reflux gently for ½ hour. Evaporate in vacuum, basify with NaOH and filter to get 8.4 g methylecgonine (V) (recrystallize-isopropanol). 4.16 g (V) and 5.7 g benzoic anhydride in 150 ml benzene with $CaCl_2$ tube to exclude water and gently reflux four hours. Cool in ice bath, acidify with HCl and dry, evaporate in vacuum (or extract with ether, basify with NaOH, saturate with K carbonate and extract with $CHCl_3$:dry and evaporate in vacuum) to get 6 g red oil which precipitates (VI) with addition of a little isopropanol.

Method 2 JOC 22,1391(1957)

Suspend 40 g beta-ketoglutaric acid in a mixture of 60 ml glacial acetic acid, 43 ml acetic anhydride and stir three hours at 10°. Filter, wash precipitate with benzene and dry in vacuum over KOH about two hours to get 30 g beta-ketoglutaric anhydride (I). Dissolve 13.5 g (I) in 50 ml cold methanol and let stand one hour at room temperature. Add this solution to a solution of 10 g methyl-amine.HCl, 4 g NaOH in 850 ml water and stir in 125 ml 0.8N succindialdehyde (preparation below). Let stand twenty-four hours at room temperature, take pH to 4 with 6N HCl and wash with 35 ml $CHCl_3$. Dry and evaporate in vacuum (or basify with 20 ml 4N NaOH and 4 g $KHCO_3$, extract with 9X100 ml $CHCl_3$ and dry, evaporate 75% of the $CHCl_3$ on steam bath, then evaporate in vacuum) to get 16 g yellow oily 2-carbomethoxytropinone (methyl tropan-3-one-2-carboxylate). Recrystallize by dissolving the oil in 30 ml hot methyl acetate and add 4 ml cold water and 4 ml acetone; let stand three hours at 0°, filter and wash precipitate with cold methyl acetate. This product is identical with (IV) of method 1 and is converted to cocaine as already described.

Succindialdehyde JOC 22,1390(1957)

Suspend 23.2 g succinaldoxime powder in 410 ml 1N sulfuric acid and add dropwise with stirring at about 0° a solution of 27.6 g Na nitrite in 250 ml water over three hours. Stir two hours at room temperature (keeping air out), stir in 5 g Ba carbonate and filter. The succindialdehyde should then be extracted from the basic solution with ether and the ether dried and evaporated in vacuum. For succinaldoxime preparation, see JOC 21,644(1956) and JACS 68,1608(1946).

Method 3 JOC 22,1389(1957)

Add with stirring over 1½ hours 192 g powdered anhydrous citric acid in 32 g portions to 202 ml (383 g) fuming sulfuric acid (21%). Make the first two additions at 0° carefully; the other four at 15°. Stir one hour at room temperature, and for three hours at 35° and 17 hours at 25°. Add dropwise with stirring below 0°, 500 ml methanol over three hours. Keep about fifteen hours at room temperature and add to a stirred mixture of 700 g NaHCO₃, 500 g ice and 200 ml water. Filter, wash precipitate with 150 ml 50% aqueous methanol and extract the filtrate with 7X400 ml ether. Dry and evaporate in vacuum (can distill 85/1) to get 110 g oily dimethyl-beta-ketoglutarate (I). Use this in method 1 or as follows. Dissolve 33.6 g KOH in 150 ml methanol and add dropwise at 0° over ½ hour (or at room temperature over one hour) to 43.5 ml (I) in 10 ml methanol. Let stand three hours at room temperature, add 50 ml ether and refrigerate twelve hours to precipitate the dipotassium salt of monomethyl-beta-ketoglutarate (II). Dissolve 10 g succindialdehyde in 200 ml water at -5° and add 41 g (II) and 11.8 g methylamine.HCl. Let stand a few hours at room temperature and proceed as in method 2.

Method 4 JOC 22,1389(1957)

Mixture of 1.35 g Na methoxide (Na in methanol), 3.48 g tropinone (which can be obtained by K dichromate oxidation of tropine), 4 ml dimethylcarbonate and 10 ml toluene. Reflux ½ hour, cool to 0° and add 15 ml water containing 2.5 g NH₄Cl. Extract with 4X50 ml CHCl₃, dry and evaporate in vacuum and dissolve the oil in 100 ml ether. Wash two times with a mixture of 6 ml saturated aqueous K carbonate and 3 ml 3N KOH (dry and evaporate in vacuum the ether to recover unreacted tropinone). Take up the oil which separates in saturated aqueous NH₄Cl and extract it with CHCl₃. Dry and evaporate in vacuum to get an oil which is dissolved in hot acetone. Cool, add a little water and rub to start precipitation of 1.5 g 2-carbomethoxytropinone. This is identical with (IV) of method 1, and can be recrystallized and converted to cocaine as already described.

Other References

The various cocaine precursors and analogs seem not to have been tested for psychedelic activity. Cogentin (benzo-tropane) is hallucinogenic at a dose of about 4 mg. Synthetic compounds which may have activity are *l*-pseudo-cocaine, tropacocaine, eccaine, eucaine, and benzoyl-N-methyl-granatoline (se BER 51,235(1918), Chem. Zentr. 1402(1939), JCS 41(1924), 1150(1925), 1429 (1932), 1511(1933), Q.J. Pharm. Pharmcol. 7,46(1934)). For a simple synthesis of pseudopelletierine, which can be reduced and acylated to give active compounds, see J. Pharm. Pharmacol. 22 (supplement),29(1970). For a review of tropane chemistry see Manske (Ed.), The Alkaloids 1,271(1951), 5,211(1955), 9,269 (1967) 13,351(1971). JCS 3575(1957), Acta Pharm. Suecica 7,239(1970) and JMC 16,1260(1973) contain further information on cocaine analogs. For Cogentin synthesis see CA 47,2218h-(1953).

It may be possible to partially oxidize butanediol to butanedial (succindialdehyde) and distill this as the reaction occurs.

Cocaine Manufacture and the Cocaine Traffic

(The following summary of cocaine preparation and trade was prepared by the California State Attorney General in 1973.)

The coca plant is an evergreen, native to South America, particularly the countries of Peru, Bolivia, Brazil, Chile and Columbia, and should not be confused with the cocoa plant, from which chocolate is made. Although the coca plant is natural to South America, it has been successfully cultivated in Java, West Indies, India and Australia.

The coca plant is grown on mountain slopes or terrace uplands that have a tropical or semi-tropical climate. Actually, the plant is grown under conditions which are little suited for other crops. These mountainous areas of South America vary in altitude from 1,000 to 6,000 feet above sea level and temperatures of 68 to 86 degrees Fahrenheit. The most suitable conditions for the development of the coca plant are clay type soils, rich in humus and iron content, situated in sheltered upland valleys and exposed to constant humidity and rain precipitation. Under ideal conditions, the plant can survive, for a century or more, growing steadily in strength. In the cultivation of the coca plant, the seeds are usually taken from a plant three or more years old. The seeds are placed in containers and germinated in damp sheltered nurseries. The seeds are watered heavily for five days until they begin to swell after which they are planted in a mixture of humus sand and earth, in equal proportions, shaded and abundantly watered. After about a week and a half, the shoots begin to appear and the germinated seeds can be transplanted within 2 months. The sparsely leaved plant is usually 6 to 10 inches in height and is transplanted in the open since it has become resistant to most climatic variations. The young plant is usually planted in small trenches varying in density from 1 to 4 plants per square yard.

Once the young plant has been transplanted (usually in the wintertime) into open fields, there is very little the cultivator does except leave the plants to themselves. Where dampness is constant and rain is regular, even irrigation is unnecessary. After approximately 1 year from the transplanting, the coca plant yields its first crop of leaves which is generally the reason the coca bush has been cultivated in the first place. The plant normally yields 4 crops of leaves per year. The coca plant is a shrub-like bush which grows from 5 to 10 feet tall with widely branched trunks containing twigs which become densely populated with leaves toward the ends. The green smooth edged leaves vary from 1 to 3 inches long and smell very similar to tea leaves. Normally, in order for the harvesting of the coca leaf to be profitable, there must be a minimum of 72,000 plants for every 10,000 square meters and the plant must last for over 30 years. Approximately 10 million coca plants produce 700,000 kilograms of coca leaf.

The average coca leaf contains from ½ to 1% of the alkaloid cocaine although there are various factors influencing the cocaine content including atmospheric conditions, age and condition of the plant, quality of the soil, fertilizers used, timing of cultivation and harvesting, the drying process, etc. It is estimated that one man can harvest approximately 30 kilograms of leaves in a day.

The drying process is very important and takes approximately 2 days of at least 3 hours of daily sunshine. During the process, the leaves must be turned over for even drying. If the drying is too extensive, the leaves will become too dry and lose their commercial value. In the drying process, the coca leaf loses more than 75% of its original weight. The leaf is divided into 3 basic categories:

1. Dark green colored leaf, dried by mechanical means or by airing and pressed into bales. This form is best suited for use in export.

2. Dark colored leaf resulting from defective drying and deliberately beat in order to meet wide demand for consumption.

3. Leaves which as a result of neglect, dampness, delay in drying or disease, have lost some of their alkaloid content, and are of practically no commercial value. This type of

coca leaf is often used as a local fertilizer.

Once the leaves are dry, they are pressed and wrapped into packages of 50 or 30 kilograms. In 1965, 1 kilogram of dried coca leaf, in Bolivia, was valued at approximately one dollar.

The coca leaf is commonly chewed by the natives of South America. The natives claim that the cocaine depresses their hunger and increases their strength. The leaves are very bitter when chewed and are often flavored with another substance such as lime. It has been estimated that over 90% of the Indians chew the coca leaf. The native chews, on an average, about two ounces of coca leaf daily and is often characterized by blackish red deposits on his teeth.

The coca leaf is either consumed by the natives of South America or exported to other countries for consumption. Another use of the coca leaf is in the extraction of cocaine either for illegitimate or legitimate use. The majority of the legal and/or clandestine cocaine factories are in South America due to the cost and bulk of transporting the whole leaf. In 1961, Bolivia produced an annual crop of from 12,000 to 18,000 tons of leaves although only half reached the legal market. The alkaloid cocaine is extracted from the coca leaf in basically three different chemical procedures. These procedures are used both in licit and illicit labs in the production of cocaine.

Manufacture of Cocaine

According to a chemist who assisted in the legal manufacture of cocaine, there are three basic methods of extracting cocaine from the coca leaf:

1. The dried coca leaf is treated, through a chemical process, with an acid solution such as sulfuric acid, producing raw cocaine or coca paste. The coca paste which contains approximately 70% cocaine, is put through another chemical process with hydrochloric acid creating a hydrochloric salt or cocaine hydrochloride which is soluble in water. This particular process is very time consuming and can take from 1 to 2 weeks to complete. This process is used by both the legitimate and illicit manufacturers of cocaine.

2. The dry coca leaf is treated in a chemical process with a basic solution such as sodium carbonate, producing raw cocaine. The raw cocaine is then put through another chemical process with hydrochloric acid creating a hydrochloric salt or cocaine hydrochloride. This process is less

time consuming than process No. 1 and it is probably the one preferred by the illicit manufacturer.

3. The third process is more advanced and technical than the other two procedures. The basic advantage of this process is that it gives a greater yield. The dried coca leaf, with its various alkaloids including cocaine, is broken down into ecgonine which is the chemical base or core of the cocaine molecule. The ecgonine is then treated with methyl iodide and benzoic anhydride in a chemical process creating pure cocaine.

Cocaine Synthesis

Coca Leaves
↓
Ecgonine
↓
Cocaine

$$H_2C-CH \longrightarrow CHCOOH$$
$$| \quad | \quad |$$
$$NCH^3 \quad CHOH \quad + \quad CH^3I \quad + \quad C^6H^5COOOCH^5C^6 \rightarrow$$
$$| \quad | \quad |$$
$$H_2C-CH \longrightarrow CH^2$$

$$H_2C-CH \longrightarrow CHCOOCH^3$$
$$| \quad | \quad |$$
$$NCH^3 \quad CHOOCC^3H^5$$
$$| \quad | \quad |$$
$$H_2C-CH \longrightarrow CH^2$$

Ecgonine   +   Methyl Iodide   +   Benzoic Anhydride ⟶ COCAINE

The Peruvian coca leaves, because of their richness, are commonly used in the extraction process as described in 1 or 2. When the dried coca leaves have a low cocaine content, the ecgonine process is preferred. Normally, it takes approximately 100 pounds of dried leaves to produce one pound of cocaine.

A chemist from the Federal Bureau of Narcotics and Dangerous Drugs, who was in Bolivia to observe clandestine cocaine operations, related the following step-by-step procedure for manufacturing cocaine. The method can be conveniently divided into three major steps: (1) extraction of cocaine from the leaf and chemical conversion to the sulfate; (2) treatment of cocaine sulfate with potassium permanganate and conversion to the free base (aka paste); and (3) conversion of the paste or free base to cocaine hydrochloride. In general, steps (1) and (2) are carried out in "sulfate" labs while step (3) is performed in "crystal" labs.

Coca leaves to cocaine sulfate

Step 1:

147

(a) A sufficient volume of warm water is used to dissolve 26 kgm of potassium carbonate. This solution is poured into a "cut-out" drum containing 250 pounds of dried coca leaves (the volume of solution is just enough to cover the leaves).

(b) The coca leaves and carbonate solution are treated in one of three ways:

1. The mixture is stirred by hand
2. The mixture is stepped on by the local Indians
3. The mixture remains untouched

(the first two methods generally take one day if done by strong individuals while the third method involves a period of four days; the treatment of the leaves with an aqueous solution of potassium carbonate allows penetration of the solution into the leaves converting any cocaine salts to the free base, allowing for subsequent extraction into kerosene).

(c) To the potassium carbonate solution and the leaves are added 200-400 L kerosene. The result of this is a greenish viscous liquid of about 300-400 L. Water and kerosene are immiscible but apparently "bridging" materials are extracted from the leaf which allows miscibility between the potassium carbonate solution and the kerosene; it was also explained that what we know as kerosene is also called benzin and paraffin, depending who is manufacturing the cocaine; this explanation clarified pre-existing conflicts about certain method terminology.

(d) The kerosene extract is separated from the leaves by draining through a plug in the bottom of the drum. The kerosene extract (300-400 L) is placed in another container and 1 L of concentrated sulfuric acid is added very slowly. A precipitate begins to come out of solution and settle to the bottom of the container (apparently kerosene-insoluble cocaine sulfate and other alkaloidal sulfates are formed).

(e) The liquid is separated from the sulfate precipitate and may be used to extract the next batch of fresh leaves (the liquid would probably have to be reconstituted, though, with potassium carbonate, etc.).

(f) The sulfate residue, about 1 kgm, is allowed to dry in the sun for about one day.

Cocaine sulfate to cocaine base (paste)

Step 2:

(a) Into a container holding 6 L of water and 360 cc of concentrated sulfuric acid the sulfate precipitate from the preceeding step is added and the solution stirred.

(b) In another container 1 L of water and 1 kgm of commercial potassium permanganate are mixed.

(c) The permanganate solution of (b) is slowly poured into the sulfuric acid solution of (a).

(d) The resultant purple-colored solution is then filtered through paper; if the filtrate is colored it is passed back through the filter (the use of potassium permanganate in the manufacturing process has been well-established; it is probably added as a decolorizing agent, with most of the colored residue remaining on the filter paper; commercial potassium permanganate is easily obtained in LaPaz through pharmacies; it is used by the local inhabitants to bathe the feet).

(e) To filtrate of about 6 L is added 1 L of ammonia; a precipitate is formed which is dried in the sun or under ultraviolet light (the resultant precipitate is cocaine free base or paste plus impurities).

Cocaine base (paste) to cocaine hydrochloride
Step 3:

(a) To the approximately 1 kgm of paste in a container is added 10 L of acetone and the resultant solution filtered through paper (the residue on the paper is probably inorganic in nature; ether can be used instead of acetone with some modification in the procedure; both ether and acetone can solubilize cocaine base while they are both good solvents for the crystallization of the hydrochloride salt).

(b) To the filtrate is added 10 L more of acetone; the resultant 20 L is passed through new filter paper. The temperature of the acetone should be about 15° C. in order to accept the hydrochloric acid and alcohol.

(c) To the 20 L of acetone are added 300 cc of concentrated hydrochloric acid and 300 cc of absolute ethanol; on the addition of absolute ethanol cocaine hydrochloride starts crystallizing out.

(d) After 3-4 hours crystal formation is complete and the cocaine hydrochloride crystals are collected on filter paper and dried in air.

Kerosene, which is used in great volumes in the initial extraction procedure is not controlled in Bolivia and is, in fact, used by the populace for heating their homes. It would theoretically be very easy to obtain kerosene in large quantities through pharmacies in La Paz. The inorganic reagents such as sulfuric acid, hydrochloric acid, ammonia, and potassium carbonate are not controlled. Ether and absolute alcohol, on the other hand, are controlled substances. Diluted alcohol (about 50%) is not controlled though and, accordingly, can be converted to absolute alcohol.

Trafficking in Cocaine

A. Legal

Once the cocaine has been legally produced from the coca leaf, it is exported to various countries for medicinal use, basically as a topical local anesthetic (applied to the surface, not injected, only treating a particular area). In the United States the crystalline powder is imported to pharmaceutical companies who process and package the cocaine for medical use. Merck Pharmaceutical Company and Mallinckrodt Chemical Works distribute cocaine in crystalline form (Hydrochloride Salt) in dark colored glass bottles to pharmacies and hospitals throughout the United States. Cocaine, in the alkaloid form (base drug containing no additives such as hydrochloride in the crystalline form) is rarely used for medicinal purposes. Cocaine hydrochloride crystals or flakes come in 1/8, 1/4 and 1 ounce bottles from the manufacturer and has a wholesale price of approximately $20 to $25 per ounce (100% pure).

Cocaine is still a drug of choice among many physicians as a topical local anesthetic because the drug has vasoconstrictive qualities (shrinks and stops the flow of blood). Synthetic local anesthetics such as novacaine and xylocaine (lidocaine) have also been discovered and used extensively as a local anesthetic.

B. Illegal

Hierarchy Of Cocaine Traffic

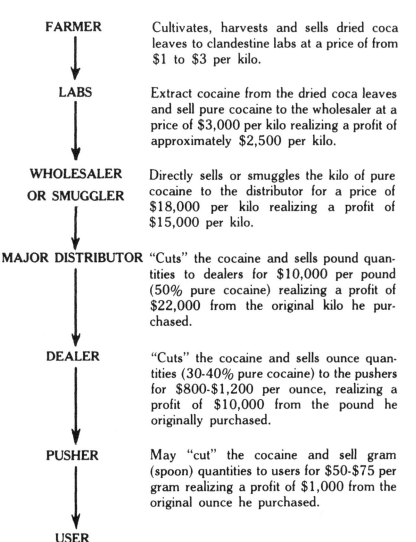

FARMER — Cultivates, harvests and sells dried coca leaves to clandestine labs at a price of from $1 to $3 per kilo.

LABS — Extract cocaine from the dried coca leaves and sell pure cocaine to the wholesaler at a price of $3,000 per kilo realizing a profit of approximately $2,500 per kilo.

WHOLESALER OR SMUGGLER — Directly sells or smuggles the kilo of pure cocaine to the distributor for a price of $18,000 per kilo realizing a profit of $15,000 per kilo.

MAJOR DISTRIBUTOR — "Cuts" the cocaine and sells pound quantities to dealers for $10,000 per pound (50% pure cocaine) realizing a profit of $22,000 from the original kilo he purchased.

DEALER — "Cuts" the cocaine and sells ounce quantities (30-40% pure cocaine) to the pushers for $800-$1,200 per ounce, realizing a profit of $10,000 from the pound he originally purchased.

PUSHER — May "cut" the cocaine and sell gram (spoon) quantities to users for $50-$75 per gram realizing a profit of $1,000 from the original ounce he purchased.

USER

The original 100 kilos of dried coca leaf that it takes to produce 1 kilo of pure cocaine costs approximately $200. The kilo of pure cocaine will eventually be worth over $200,000 when sold to users in 25% pure gram quantities.

The illicitly manufactured cocaine from the various clandestine cocaine labs in South America, is smuggled to various countries including the United States for black market trafficking and use. Those involved in the smuggling of cocaine vary from a one man operation to organized syndicates. The smuggling methods are unlimited and vary with one's imagination. Often times, cocaine is first smuggled into Mexico rather than directly from South America to the United States. Cocaine, as is heroin, is usually packaged in hermetically sealed plastic bags or rubber condoms for smuggling purposes. Once the cocaine enters the US, it is then distributed through various sub-dealers down to the users. Illicit cocaine, basically, comes in three forms:

1. The hard tiny rock form which is readily available, especially to the large wholesaler or dealer.

2. The flake form which is generally fairly pure cocaine which has been broken down into tiny flakes and considered a delicacy among users of cocaine.

3. The powdered form which is usually rock or flaked cocaine diluted with other substances such as lactose or procaine.

In the illicit traffic of cocaine, as in many other drugs, there is a definite channel which the drug goes through from the harvester to the user. Initially, there is the farmer who cultivates, dries and ships the coca leaf to the illicit lab. The clandestine labs then chemically extract 90 to 100% pure cocaine from the leaf. From the lab, the cocaine is usually sold to smugglers or wholesalers at a price of $200 an ounce or $3,000 a kilo. The wholesaler smuggles the cocaine into the United States and sells it to a major cocaine distributor for a certain agreed upon price which varies and ranges from $18,000 to $22,000 a kilo. The distributor will then take the large quantity of cocaine and sell lesser amounts to a number of dealers. He may sell the cocaine in its pure form or dilute it and sell more for a lower price. Most of the traffickers keep in mind that cocaine loses its strength readily and sometimes the cutting or diluting agent will have a tendency after a period of time to begin destroying the cocaine content. When the dealers are in possesison of their ½ pound or pound of cocaine, they will most often dilute it with a cutting agent and sell it in ounce quantities to even smaller dealers. The cocaine street pusher will in turn, sell it to the user in gram quantities.

The paraphernalia and diluting agents for the cutting of cocaine are very similar to those used for heroin. One of the basic differences between "stepping on" (diluting) cocaine as compared to heroin, is that cocaine is usually only diluted down from 20 to 40%. The process for cutting cocaine varies from individual to individual with often times the large dealer using a more elaborate process, but the basic operation is the same throughout cocaine traffic. The basic paraphernalia used is:

1. Diluting agent
2. Scales
3. Measuring spoons
4. Flat nonporous surface
5. Razor blade, playing card or some other sharp-edged instrument
6. Sifter or nylon stocking
7. Funnel
8. Packaging container such as rubber condoms, tinfoil bindles or plastic baggies

The cutting or diluting agent used for cocaine again varies with the individual and the substance that is readily available to that individual. Some of the common cutting agents for cocaine are:

1. Procaine which is a synthetic preparation in powder form used as a local anesthetic.
2. Mannite, a sugar substance used as a laxative produced in Italy.
3. Menita, a milk sugar from Mexico and South America.
4. Lactose or Dextrose, a white powdered milk sugar used as a baby food supplement and purchased readily in the United States in any drug store.
5. Powdered methamphetamine also known as speed.
6. Epsom salts.
7. Quinine used to treat leg cramps and malaria.
8. Powder vitamins purchased in health food stores.
9. Just about any soluble powder that is not disruptive to the body can be used, such as baking soda, powdered sugar, powdered milk, starch, etc.

The dealer will either be told the percentage of cocaine by a trusted "connection" or he will be able to approximate the percentage by various means. Some of the ways of ascertaining the approximate percentage of cocaine and the cutting agent are:

1. *Quantitative chemical analysis* which is an elaborate process requiring a qualified chemist and some elaborate laboratory equipment.

2. *Cocaine drug testing kits* either manufactured for law enforcement purposes or produced by the underground. These testing kits are simply presumptive color tests. The basic color test used for cocaine is cobalt thiocyanate. The cocaine or any of the other substances from the caine family will form a brilliant blue flaky precipitate in the cobalt thiocyanate. This is an indication that the product is cocaine, procaine, tetracaine, etc. In order to determine whether there is actually any cocaine and not all procaine, stannous chloride is added to the precipitate causing all of the caines except cocaine to dissolve. If the dealer suspects that the cocaine has been cut with another caine, he can then make a partial determination as to how much of the procaine or other caine is contained in the total powder.

3. *Chlorox test.* It is alleged that the dealer can take suspected cocaine and drop it in a vial of clorox. Presumably the cocaine will dissolve completely and procaine will turn a reddish orange color with any other cut trailing to the bottom of the vial as residue.

4. *Water Test.* It is also alleged by the street dealers, that a determination can be made as to how much cut is in the cocaine by placing the powdered substance in a glass of water. The cocaine will dissolve almost immediately leaving the remaining cut which normally will dissolve slower and not as clear.

5. *Burning test.* The powdered cocaine is placed on aluminum foil and held over a low flame or match. The cocaine will burn clear. A sugar cut will darken and burn a dark brown or black therefore the larger the cut, the darker the burn. Crystallized speed or methamphetamine will pop when burned. Salts do not burn and remain as residue (cuts such as procaine or quinine also burn fairly pure although it is alleged that procaine can be detected by a bubbling of the substance before it burns clear).

6. *Methanol test.* Most common cuts do not dissolve in pure alcohol although cocaine does. Unfortunately for the dealer, procaine and methamphetamine also dissolve in pure alcohol. It is imperative that pure methanol be used since any water in the alcohol will tend to dissolve sugar and salt. Methanol can be obtained in most paint supply stores as methalated spirits. The dealer will take two equal amounts of the cocaine substance and place the equal amounts in two teaspoons next to one another. At this time, ¼ of a teaspoon of pure methanol is added to one of the spoons. The mixture is then stirred and any powder that remains is compared to the original unaltered amount in the second teaspoon to determine the percentage of the cut. If, for example, 20% of the original amount did not dissolve, the substance tested would be no more than 80% pure. If the suspected cut is procaine, the cocaine substance can be added to sodium carbonate solution. This would dissolve all the cocaine leaving just the procaine.

7. *Use test.* Some dealers will test the percentage of cocaine by inhaling (snorting) it into the nostrils. This is probably the best and most common street test in determining the purity of the cocaine. The tester should standardize the amount snorted so that he will have the ability to distinguish. The tester will look for the swiftness of the high and the "freeze" or numbness the substance causes. If the nasal passages burn and the eyes tear, there is a good possibility the cocaine has been cut with speed. Sugar and salt cuts will often times cause a post nasal drip. Excessive sweating and hyperactivity could mean either a speed or quinine cut was used. Excessive diarrhea would denote a laxative type cut such as epsom salts or menita. Speed tends to cause irregular bowel movement. A greater degree of numbness indicates the presence of procaine or other local anesthesia.

8. *Taste test.* Cocaine has a bitter taste and the addition of any cut will tend to alter that taste. A milk sugar cut will sweeten the cocaine although dextrose has a tendency to sweeten the substance more than lactose. Procaine will be bitter to the taste but will tend to numb the gums and tongue quicker and longer than cocaine. Salt has an after taste and epsom salts are a bit more sour in taste and sandy in texture.

9. *Observation test.* Pure cocaine crystals have a shiny almost transparent appearance and even when crushed, will retain the

crystalline sparkle. The crystalline sparkle of cocaine will be dulled by most cuts. Dextrose has less dulling effect than lactose although a speed cut usually dulls the crystals less than most other cuts. Although salts have a crystalline structure, they tend to be duller than the cocaine crystals. An alleged indication of the purity of the cocaine is the tiny rock-like material contained within the total substance. The tiny rocks are allegedly pure cocaine as they come from the manufacturer. The rock or hard substance can be felt by feeling the powdered substance.

Once the dealer has ascertained the purity of the cocaine and/or the cut or diluting agent used, he is then ready to begin the process of "stepping on" the cocaine. Most dealers will dilute a small portion of the cocaine and then re-test it. Most of the dealers claim that they usually only cut the amount of cocaine that will immediately be sold due to the fact that the cuts have a tendency to destroy the stability of cocaine. It is therefore advantageous to the dealer to keep the cocaine sealed in a cool place such as the refrigerator and in an amber or dark-colored jar to retain the strength of the drug as long as possible. Dealers claim that with time, moisture, warmth, air and sunlight tend to decrease the potency of the cocaine.

The process for "stepping on coke" again varies with individuals but the two basic formulas are similar to those of heroin and are as follows:

1 ounce of Lactose added to 1 ounce of 100% Cocaine = 2 ounces of 50% Cocaine
2 ounces of Lactose added to 2 ounces of 50% Cocaine = 4 ounces of 25% Cocaine
2 oz. Lactose added to 1 oz. 100% Cocaine = 3 oz. of 33 1/3% Cocaine
3 oz. Lactose added to 1 oz. 100% Cocaine = 4 oz. of 25% Cocaine
4 oz. Lactose added to 1 oz. 100% Cocaine = 5 oz. 20% Cocaine

The dealer will measure out the desired amount of cocaine, for instance five level teaspoons, and place it in a pile on a flat nonporous surface such as a record album, mirror or glass plate. He will then measure out the desired amount of lactose and place it in a separate pile on the same surface. Then, using a playing card, razor blade, knife or any sharp edged instrument, the dealer chops the cocaine to take out all the lumps so the cocaine is a fairly fine powder. The cocaine is then sifted through a sifter or nylon stocking producing a fine fluffed powder and removing foreign material from the substance. Once through the sifter, the cocaine usually has a little more volume since it has been fluffed. The cocaine is sifted into a pile and the same process is repeated with the diluting agent. The dealer will then mix the pile of cocaine into the pile of diluting agent. Once this has been accomplished, he sifts the diluted cocaine through a sifter trying to get the mixture as equal as possible. The dealer may resift the diluted cocaine to assure an equally distributed mixture. The sifted cocaine is placed in a single pile. At this time, the dealer is ready to place the cocaine into packages for sale.

In larger quantities, the cocaine is usually packaged in either airtight rubber condoms or plastic baggies that are also airtight. A rubber condom will usually hold from ½ ounce to 3 ounces of cocaine whereas a plastic baggie can hold up to 6 ounces of cocaine. It appears that most dealers prefer using the plastic baggie since it is believed that synthetic hard plastics and rubber tend to react unfavorably with cocaine in terms of chemical composition and therefore in time, decrease the potency of the cocaine. Smaller amounts or gram quantities of cocaine are usually packaged in smaller airtight clear plastic bags, paper bindles or tin foil bindles. It is very uncommon to see cocaine packaged in toy balloons the way heroin is packaged because of the action of the rubber with cocaine.

Once the dealer has diluted the cocaine, he will measure out the desired amount to be packaged. This can be done by measuring it on a scale.

For ounce quantities, the dealer will measure on a scale approximately 28 grams and then place this powder into a rubber condom or a plastic baggie. The dealer will then, using a rubber condom, tie a knot in the condom and possibly fold the open end back over for added protection. When using a plastic baggie, the dealer will get as much of the air out as possible and either fold the baggie over and seal it with scotch tape or use a twister to seal the end of the baggie. Another method of determining the proper amount of cocaine to be packaged is by using measuring spoons. If the dealer is going to use the most common measure of cocaine, that is a "spoon" or approximately a gram, he will measure out a level half teaspoon and place it in the proper packaging device. When using paper bindles, the dealer will place the cocaine on the extreme inside of the paper bindle, usually a square piece of paper with 4 inches by 4 inches being a common size and drawing the two opposite ends together to form a triangle. The dealer will then begin at the base of the triangle and fold approximately ¼ inch folds in the paper bindle until there is approximately an inch of unfolded triangle left at the top. The outer wings of the paper bindle are then folded inside and the unfolded top of the bindle is folded and tucked into the wings. When using tin foil, the dealer will put the desired amount inside a square piece of tin foil, fold the square over into a rectangle, seal the top with a small fold and then fold the extreme ends into the inside.

Once the cocaine has been packaged, it is ready to sell to either smaller dealers or to users.

How to Grow Coca Plants

1. Seeds should be planted as soon as they fall from the bush. If they dry out, they will die right away. The only way to keep them for a maximum of about two weeks, is to keep them in moist (not wet) sphagnum in a cool place. Often this initiates germination, so they must be watched for rot or premature germination. Under no circumstances should they be kept dry, since even room humidity is too dry.

2. Vermiculite seems to be the best medium for coca germination, fine grade if possible. Styrofoam cups are OK, but I prefer small plastic pots, 2"diameter, with holes in the bottom. Seeds should be planted no deeper than one inch. Pots should be raised so as not to saturate the medium. Coca, whether as a seedling or a mature plant, never likes to have wet feet. I think it is better to start them in small pots rather than flats, so there is less damage to the root system when they are transplanted. Forget the hot pad -- I think it is completely unnecessary. Seedlings usually come up in 2 to 4 weeks if they are viable.

3. Since most people don't have enough room in their shower stalls for plants, I'd say forget this one, too. Seeds will germinate in any warm place, even if the humidity is not too great. A better idea is to place your germination pots in a terrarium with a coarse gravel layer on the bottom. Do not seal over and allow plenty of ventilation if you choose to place a layer of glass over the terrarium. Any box of this sort will do. If possible, place a Growlux fluorescent fixture, with two 40 W bulbs, over the terrarium, especially after seeds germinate. A common problem at this state is etiolation (too little light) which makes the plantlets weak and very susceptible to damping off, a fungus attack of the tender stems.

4. Water the seeds when the vermiculite starts to dry out. Once a day is probably too often, unless you live in a very dry apartment. But if the drainage is good and you have plenty of holes in the bottom of the pots, excess water should drain off.

Fungal attack is a real problem in a humid atmosphere and another reason for keeping the plants out of your shower, a basically unhygenic place for plants.

5. Transplanting: plantlets can remain in vermiculite starting pots until they are about 2-3 inches tall. The growlights should be about a foot above the plants. I do not recommend clay pots at this stage. They dry out too fast, especially in a dry apartment. Even in one day, a fast-drying shock can kill your plants. It is better to move into plastic pots, but the size should be increased gradually. A big pot is not necessarily good for a small plant, in fact it is not a good idea at all. From styrofoam cups, I suggest a two-inch pot, then increase 1-2 inches per transplanting.

6. Soil mixture: forget the vermiculite from now on. It holds too much moisture and makes for saturated, unhealthy soil. I suggest the following: ¼ coarse clean sand, ¼ perlite, ¼ sterilized loam, and ¼ milled peat. If this seems too light, increase loam and peat. Some sterilized organic compost, screened, may also be added for nutrition.

7. Even when the plants are still in vermiculite, feeding with soluble plant food is recommended. They are heavy feeders and every three weeks or more often is not too often to fertilize. When plants are older it is important to give them iron in the form of iron chelate, available as a red powder sold as KEELATE on the West Coast. A yellow powder, not as good, is sold as SEQUESTRENE. This element should be added about every six months, but strictly according to instructions. Soil must be flushed three times after applying the dissolved iron compound to avoid burning roots. Most yellowed or bleached out leaves are caused by iron deficiency, but this also occurs when plants go deciduous. Periodically, the whole coca bush turns yellow and drops its leaves, every one. Most people freak out when this happens, but if it is otherwise a healthy, vigorous plant, then this is normal. After dropping, new flushes soon appear to renew the foliage. This is more likely to happen with *Erythroxylum coca* than with *E. novogranatense*.

8. Transplanting depends on the size of the plant and how fast it is growing. If you think your plant needs transplanting, look at the holes in the bottom of the pot to see if any roots are present. If so, then the roots have probably filled the pot and it is time. You can also carefully de-pot the plant by tapping upside down

on a table edge. Repotting is probably unnecessary unless the roots have encircled the inner periphery of the pot. Again, the size of the pot should be increased gradually for best growth.

9. Watering: most city water, is unsuitable for coca. They are calciphobes and don't like heavy salts in the water. Best to use rainwater, melted snow, bottled spring water or distilled water if they are available. Plants should only be watered if the soil dries out. Stick your finger in the soil. If it feels moist, don't water.

10. Bugs: coca is amazingly resistant to insects and mites. Mealy bugs are the worst offenders. These may be removed with a forceps or cotton swab dipped in 50-70% alcohol. Keep infested plants in quarantine. Malathion may be used as a last resort, but then leaves cannot be used until the next flush (of leaves).

11. Light: warm, sunny exposure indoors. Full sun (through a window) will not hurt plantlets over 3 inches tall. But no full sun outdoors until they are 3 feet tall. If plants are put out in the summer, they should be protected from sun, rain, and wind, until they are large and strong. Put them in a shady place first, under a tree, etc., and gradually move to a sunnier location. Breezes are good for plants and even indoors a fan on low should be directed towards the plants. It makes them stronger.

12. Plants can also be grown entirely under growlights, or a combination of growlights and window light. Most apartments are not sunny enough for strong growth, so especially in winter, give the plants accessory light. Growlux Widespectrum Tubes seem to work well. I use one Growlux and one regular Sylvania Lifeline tube in each fixture. They work very well. The lamps are suspended 6 inches to one foot above larger plants.

13. Careful removal of the older leaves does not harm plants, but they should be strong and healthy to allow this, and probably three years old if grown indoors.

14. Coca does not like extremes of any kind. 50° F. is the lowest permissible temperature, 90° F. the highest. Sudden temperature changes are especially damaging. Likewise, sudden changes in air humidity or soil moisture. *E. novogranatense* tolerates extremes, especially droughts, better than *E. coca*, which is a much more delicate plant, but the one which

produces the most alkaloid.

15. Coca cuttings root very poorly. I have managed to root some *E. novogranatense* cuttings only after six months in perlite with an initial application of Hormodin #1 rooting hormone. It is better to fertilize your flowers and plant seed. Some varieties are self-compatible (self-fertilizing). Others require two plants of different stylar lengths (long styled x short styled) to produce seed. This is routinely accomplished by bees and other insects in the greenhouse during the summer months and can be done with a fine artists brush at home, merely by dusting pollen from flowers on one plant to those on another with opposite stylar form.

In California, outdoor cultivation of coca is possible only around San Diego, if there. Trujillo Coca would probably do well there under irrigation and intensive care. Elsewhere, forget it. I do not subscribe to growing it commercially indoors and doubt if the produce would be worthwhile. Greenhouse and apartment grown leaf is very inferior in flavor and potency. Fresh air and sunshine are in order (as with Cannabis).

Additional Notes

1. When plants are sprouting, it is OK to have several of them in the same pot -- a 5-inch clay pot will do for between 4 and 7 sprouts. When they reach at least two inches tall, it is good to transplant them into individual pots using the soil mixture recommended earlier.

2. In handling the young plants, no matter how tall or short they are, always be careful not to touch the young plants or to touch them as little as possible, particularly on the roots and on the tips of the stems. The tips of the stems are where the shoots come from that allow the plant to grow, and even when the plant is mature, never touch the end of the stems and never remove the leaves that cling precariously to the end of the stem.

3. Don't freak out when the plants go deciduous, usually about a year or a year and a half from sprouting. They drop almost all their leaves except the ones at the tip of the stems, turn yellow and mottled, and you think they're dying. They're not -- in fact, they're growing! Within a few days, little spike-shaped green sprouts will appear, and tiny, usually white, flowers. After a few years, the flowers will start producing little

seedpods, roundish oval shaped green pods that the flower may still cling to. These then dry and turn slowly red on the plant, reaching a bright red like a cherry-colored coffee fruit, which contains the albumin and nourishment for the tiny seed in the center. Usually the shrubs will go through the leaf-fall several times, about once every 2 or 3 months, before the seedpods appear. Don't expect seeds until the plant is 3 to 5 years old.

4. Back to when the plants are still sprouts. Every day -- usually in the morning, but it depends on what fits your schedule best -- once a day, flush the pots with clean water, preferably rainwater or distilled. Literally hold the whole pot (without its saucer) under very gently flowing water poured into the vermiculite or soil without touching the plant. The soil or medium should almost let the water drain straight through, retaining moisture but not water in the medium. This is the way to "water" a young plant. When they get older, you can just water them regularly like any other plant, but lightly, daily.

5. The most important thing in tending young plants is to keep the temperature *even* and constant, day and night, around 64° F. They can stand slightly higher or lower temperatures but they can't stand *shifting* temperatures.

6. Once the plants get to be above a foot, they are pretty well established. After that first scary leaf-dropping, you will learn to recognize that process when it happens as described in Note 3 above. There is a different phenomenon that looks somewhat similar that happens to plants if they go through a sudden temperature change, especially if it gets cold suddenly or if they are exposed to cold fog and winds without much warm sunlight. In this case, the leaves very quickly become dry and crinkled and turn deep brown and yellow-brown mottling, at first on their leaf tips and soon covering the whole leaf. This means your plant is about to die. The only thing to do is to lightly spray the leaves with pure (not tap) water and keep the plants at a constant warm temperature and talk to them and keep careful watch on them. Don't over-water, but keep the leaves themselves warm and moist. The plant has a 50% chance for recovery.

MISCELLANEOUS PSYCHEDELICS

Mexican Mint

Leaves of the Mexican mint Salvia divinorum are chewed by some Mexican Indians for their hallucinogenic properties. Various other species of this genus grow in the USA (e.g., the purple sage), but chewing and swallowing a handful of winter leaves of five different species growing in the western USA produced nothing other than mint-flavored burps. Either a larger amount is necessary, or only the Mexican species is active. Nurseries and seed companies sell various species. *Salvia divinorum* (hojas de la Pastora) is cultivated by the Mazatecs of Oaxaca, Mexico and seems not to grow wild. They use the juice from about sixty leaves. The chia seeds popular with natural foods people come from a species of Salvia.

Catnip

Smoking the leaves of catnip *(Nepeta cataria)* produces only mild effects in man. However, the pure active agent, cis, trans-nepetalactone, seems not to have been tested on man. Seeds are available from several companies and the freshly picked leaves are probably more potent. Catnip is not orally active in cats. For synthesis see Corsi Semin. chim. 11,93(1968) et seq.; BCSJ 22,1737(1960); Proc. Chem. Soc. 166(1963). JOC 37,3376 (1972) gives a synthesis of dihydronepetalactone (more attractive to cats than nepetalactone).

Nitrous Oxide

The laughing gas trip is interesting, but very short (a few minutes). Since breathing N_2O or N_2O-Air for longer periods will produce anoxia, mixtures containing 20% O_2 and up to 80% N_2O can be breathed for longer periods. Do not breathe gas directly from a tank, since this can freeze your lungs, and do not fit a mask tightly over the face.

LAUGHING GAS (And/Or Press, 1973) is the best reference on the psychedelic effects of nitrous oxide. N_2O can be produced by heating ammonium nitrate at 240° until gas evolution ceases (the gas should leave the flask through a tube passing successively through a water trap and a water filled bottle, before collection in a plastic bag).

Rat Root

The root of the plant *Acorus calamus* (also called flag root and sweet calomel), which grows over much of the world, is chewed by the Cree Indians of Canada to produce psychedelic effects. The active compound seems to be asarone — a precursor in an hallucinogenic amphetamine.

Ibogaine

This compound, the active constituent of *Tabernanthe iboga*, a plant used for its psychic effects by African natives, has recently been synthesized. However, there seems to be nothing in the trip it produces to justify the arduous synthesis (likewise for the structurally similar yohimbine).

Kava Kava

Extracts of the plant *Piper methysticum* have been used in Polynesia, probably for thousands of years, and *Piper plantagiveum* is similarly used in Mexico and the Caribbean. They produce a sleepy, relaxed feeling with eventual difficulty in walking. Dihydro-methysticin seems to be the most active constituent, but even this has little effect until about 3 g is taken. There would appear to be no point in synthesizing this or the other active compounds, but those interested may consult JOC 24,1829(1959) and Prog. Chem. Org. Nat. Prod. 20,131(1962), and Pacific Sci. 22,293(1968). For the best review see Bull. on Narcotics 25:59-74(1973). Some people find kava extracts quite pleasant. There is very little human research on these compounds and probably won't be (unless they become popular psychedelics).

Heimia

The leaves of various *Heimia* species (sinicuichi to the natives), found in the highlands of Central and South America, are reputedly psychedelic. One species is also found in the southern US. The hallucinations are mainly auditory and the active compound appears to be cryogenine (vertine).

Betel Nut

The nut of a palm tree is chewed by millions is Asia and elsewhere with the leaves of *Piper betel* to produce mild stimulatory effects. Arecolin and arecaidinen are among the active constituents, but they appear to be psychedelically uninteresting.

Canary Weed

The blossoms, and possibly the leaves and beans, of *Genista canarienis* and probably the related genus *Cytisus* give a mild psychedelic effect when smoked. These plants are available at many nurseries.

Club Moss

The club moss *Lycopodium gnidiodes* (known as somarona to the natives of Madagascar) is said to produce effects similar to those of marijuana when smoked. Other species of this genus are found in the USA, and the pictures are given in any botany text. Some members of the genus *Myrothamnus* are also active.

Nutmeg

This trip is best avoided since it produces delirium and other toxic effects.

Mimosa

The roots of the sensitive plants of the genus *Mimosa* are supposed to be psychoactive. They are known to contain DMT.

Phenothiazines

Some phenothiazine derivatives (tranquilizers) can be hallucinogenic at high doses (e.g., imipramine (Tofranil) at oral dose of about 1 g and Ethopropazine (Parsidol) at 100 mg).

Morphine Analogs

Some of the morphine analogs can occasionally be hallucinogenic. Nalline (Nalorphine) used to monitor heroin use, M285 (Cyprenorphine) and cyclazocine are some of the better known examples. See the Merck Index for references to the synthesis of these compounds.

Morphine Analogs — Psychopharmacologia 30,108(1973) describes a thebaine derivative, active at less than 100 micrograms, which produced a trip described as a combination of downers, LSD and itching powder.

Dextromethorphan, contained in many nonprescription cough medicines, will produce a heavy psychedelic trip, but the nausea characteristic of the opiates may constitute a problem. Kosterlitz and Villareal (Eds.) AGONIST AND ANTAGONIST ACTIONS OF NARCOTIC ANALGESIC DRUGS (1972) and Braude et al. (Eds.) NARCOTIC ANTAGONISTS (1973) are useful.

PCP

This compound (also known as Sernyl, Phencyclidine, angel dust and 1-(1-phenylcyclohexyl) piperidine) has been used as an animal tranquilizer and as a general anesthetic for human surgery. Clinical tests have shown that it has a strong tendency to produce bummers, even in people who dig other psychedelics, but under appropriate conditions some subjects like it. It produces no visual effects, a tendency to fear and anxiety, and could not easily be confused with any other psychedelics. The trip is short (about two hours) at least with low doses (about 10 mg). PCP is probably a waste of time.

The common practice of mixing PCP with LSD is stupid since they are antagonistic.

Antiparkinson Drugs

Various drugs used in the treatment of Parkinson's disease (e.g., Benactyzine) can be hallucinogenic at higher doses. However, since they seem to produce a trip like that of the glycolate esters (Ditran, etc.), which they structurally resemble, these compounds should be avoided.

Belladonna

This drug got its name from the practice of certain Renaissance ladies who used it to dilate their eyes, thus enhancing their beauty. Hyoscyamine is the active constituent of the shrub *Atropa belladonna* and of jimson weed *(Datura stramonium)*. The related drug scopolamine occurs in various plants such as henbane *(Hyoscyamus niger)*. These compounds are found in many nonprescription cold remedies, motion sickness pills, etc. The hallucinogenic effects of these plants have been known over much of the world for millennia, and have been widely used by poisoners, oracles, witches, doctors, priests, and heads. They first produce excitation, dry mouth, increase in heart rate, etc., and later hallucinations and nervous depression. The trip is *very* heavy, people frequently being delirious and hallucinating for long periods with a total loss of ability to tell the real from the imaginary; carrying on conversations with imaginary people, smoking nonexistent cigarettes, etc. These compounds are not to be recommended as psychedelics since they have a strong tendency to make you sick, delirious and totally spaced out.

Ditran and Other Glycolate Esters

There is usually complete amnesia for all but the early portion of the experience. However, judicious use of small quantities may provide a separate reality unattainable by any other means (see R. Alpert BE HERE NOW (1971) near the end of the Bhagavan Das section where he describes JB-318 (a glycolate ester)). Originally this book contained a chapter on these compounds, but after several trips, I decided they were too heavy for general consumption. A careful study of the effects in man of ditran, scopolamine and atropine has appeared (Psychopharmacologia 18,121(1973)). The study concluded that the 3 drugs are qualitatively similar and that the effects are best classified as simple delirium.

MTQ

Also known as methaqualone, Quaalude, 'ludes and wallbangers. This drug is addictive, sometimes fatal (especially when combined with alcohol) and is best left alone.

Aminotetralins

A new class of compounds with probable hallucinogenic activity is 2-amino-7-hydroxytetralins. Life Sci. 12, pt. 1,475(1973) gives references and JMC 16,804(1973) gives a synthesis of related compounds.

Ketamine

Snorting ketamine gives brief but bizarre effects. See CA 61:5569d or US Patent 3,254,124. for synthesis and the recent writings of John Lilly and others for effects.

LITERATURE AND CHEMICAL HINTS

Journal Abbreviations

ACS = Acta Chemica Scandinavica
AP = Archiv der Pharmazie
BCSJ = Bulletin of the Chemical Society of Japan
BER = Berichte der Deutsche Chemische Gesellschaft
BSC = Bulletin de la Societe Chimique de France
CA = Chemical Abstracts
CCCC = Collection of Czechoslovakian Chemical Communications
CJC = Canadian Journal of Chemistry
CPB = Chemical and Pharmaceutical Bulletin
CT = Chimie Therapeutique (chimica Therapeutica)
GCI = Gazzetta Chimica Italiana
HCA = Helvetica Chimica Acta
JACS = Journal of the American Chemical Society
JBC = Journal of Biological Chemistry
JCS = Journal of the Chemical Society
JGC = Journal of General Chemistry (English translation of Zhurnal Obschei Kimii)
JHC = Journal of Heterocyclic Chemistry
JMC = Journal of Medicinal Chemistry
JOC = Journal of Organic Chemistry
JPS = Journal of Pharmaceutical Science
LAC = Leibigs Annalen der Chemie
MON = Monatshefte fur Chemie
REC = Recueil Travaux Chemiques
TET = Tetrahedron
TL = Tetrahedron Letters

Chemical Hints

Much useful information can be found in any lab text in organic chemistry such as that of Wiberg.

Much information on chemicals, equipment, etc. can be obtained from the various trade journals and newspapers such as *Chemical Marketing Reporter*. Many companies sell used tabletting machines (e.g., Union Std. Equip. Co. in New York and Chicago and Chemical and Process Mach. Corp. in New York).

Drying

Shake the solution with an anhydrous salt such as $MgSO_4$, Na_2SO_4, etc. and filter out the salt. Solids can be dried by spreading on a filter paper at room temperature or drying in an oven at low heat.

Solvents

All solvents should be anhydrous unless otherwise specified. This can sometimes be done by drying as above and is to be attended to especially in the case of ethanol which is available in 95% or 100% (100% takes up water from the air very rapidly).

Joints

Whenever apparatus with ground glass joints is used, Dow silicone grease provides excellent lubrication and airtight seal.

Petroleum Ether

This usually refers to the light boiling fraction (60° - 80°) of petroleum ether which must not be confused with "ether" which refers to diethyl ether.

Vacuum Evaporation

This requires a heavy-walled flask. Ordinary lab vacuums are about 15 mm Hg. A simple water-forced suction vacuum requires only a water source to produce a vacuum of about 25 mm Hg, which is satisfactory for most purposes. Evaporation causes the temperature to drop which slows evaporation. Running a stream of warm water over the flask or putting it in a warm water bath avoids this. To avoid difficulties in getting residues out of the bottom of the flask, it is useful to do the evaporation in a vacuum exsiccator shaped as shown or in a flat dish in the exsiccator. Whenever a forced water vacuum is used, it is wise to place a water trap between the vacuum and the solvent being evaporated to prevent water from entering when the pressure fluctuates.

Vacuum Exsiccator

Stirring

This can be done in the old way with a stirring propeller entering through one of the necks of the flask, attached to a nonsparking motor. It is easier to sit the flask on a magnetic stirrer, and drop a magnetic stirring bar (preferably Teflon coated and egg shaped for round-bottom flask) in the solution.

Heating and Refluxing

Do not smoke or have any flames (such as pilot light on gas appliances) around when using organic solvents, especially ether. Bunsen or other types of gas burners are generally outmoded. Much better are electric heating mantles available for each size of round bottom flask. Put a rheostat in the circuit to regulate the temperature of the mantle. For refluxing, adjust the rheostat so that the vapor level in the water-cooled condenser is about a quarter of the way up the condenser. Heating plates or combination heating plate-magnetic stirrer is also useful.

Ventilation

If a lab hood is not available, some forced air ventilation such as a large fan near a window is advisable, especially if ether is used.

Lithium Aluminum Hydride

Lithium aluminum hydride is expensive and often difficult to obtain. It can be synthesized (JACS 69,1197(1947)), but this is rather tricky. Its use can usually be circumvented by using a different reducing method or a different synthetic route.

Safety

An explosion shield, asbestos gloves, face mask and tongs are desirable.

A substitute for Raney-Nickel, JACS 85,1004(1963)

For the use of $NaBH_4$-Nickel in hydrogenation see JOC 38,2226(1973). For various articles on hydrogenation see Ann. N.Y. Acad. Sci. 214(1973).

5 mMoles Ni acetate in 50 ml water in 125 ml Erlenmeyer flask; connect as below to Hg pressure outlet. Flush with N_2, stir and add over 30 seconds with the syringe, 10 ml 1M $NaBH_4$ in water. After hydrogen evolution ceases, add 5 ml more of $NaBH_4$. Decant the aqueous phase, wash solid with 2X50 ml ethanol to get the Ni-boride catalyst as a black, granular solid. Hydrogenation can then be done as described below at room temperature and atmospheric pressure.

A Highly Active Raney-Nickel Catalyst, JOC 26,1625(1961)

Add with stirring, in small portions over ½ hour, 40 g 50% Raney-Ni to 600 ml 10% NaOH in a 1 L three-neck flask and continue stirring one hour. Let the Ni settle and decant the solution. Wash residue with 5X200 ml water, 5X50 ml ethanol, always keeping the Ni covered with liquid. Store under ethanol in refrigerator. Hydrogenation with this catalyst can be carried out in a low pressure Parr bottle (e.g., 30-80 ml ethanol, 5-10 g Ni suspension, 1-2 ml 20% NaOH, 40° - 50° and 40-60 PSI H_2).

A New Method for Hydrogenating at Room Temperature and Atmospheric Pressure

Method 1: External Generation of Hydrogen JACS 75,215(1953)

This method reduces the solvent volume in the reducing flask for large-scale work. Add 1M $NaBH_4$ in water to an aqueous HCl or acetic acid solution containing a little $CoCl_2$, if necessary for a rapid rate.

Method 2: Internal Generation of Hydrogen JACS 84,1494-5, 2827-30(1962)

Three-necked flask fitted with a graduated dropping funnel or a 50 ml burette, an inlet port fitted with a rubber serum cap, and an Hg manometer which allows gas to escape when the pressure exceeds about 25 mm above atmospheric pressure. At room temperature (25° water bath) with stirring, add 1 ml 0.2M chlorplatinic acid (commercial 10% is about 0.2M) to 40 ml ethanol and 1 g decolorizing carbon (e.g., Darco KB — may omit this but the reaction is slower). Flush with N_2 if possible and add 5 ml 1M $NaBH_4$ (prepared from 3.8 g $NaBH_4$, 95 ml ethanol, 5 ml 2N NaOH) rapidly to a vigorously stirred solution (black precipitate forms). After about one minute inject about 4 ml concentrated HCl or glacial acetic acid to initiate hydrogen generation. Then inject about 0.02M of the unsaturated compound. Add $NaBH_4$ dropwise so that flask pressure remains about atmospheric pressure. Reaction is over when uptake ceases. $NaBH_4$ addition can be made automatic by putting a syringe full of it in an Hg-filled tube (about 15 mm Hg, with holes so that as the pressure drops, more solution enters). This apparatus is available from Delmar Scientific Labs. Can also use $PdCl_2$, $PtCl_4$, Ni, Rhodium, PtO_2, platinum-carbon catalyst, palladium-carbon catalyst in place of shloroplatinic acid, but the last three are poor. Other salts can be used in place of the chlorides.

Note that the above procedures permit hydrogenation without the use of hydrogen tanks or special hydrogenation apparatus (Parr bottles, etc.).

Other references on the use of boron compounds for reduction: Org. Rxns. 13,28(1963); JACS 86,3566(1964); JOC 28,3261 (1963); JCS 371(1962).

Crystallization

Extreme dryness of solvent is often necessary for crystallization. This can often be accomplished (e.g. for diethyl ether) by carefully adding lithium aluminum hydride to the predried solvent, mixing thoroughly to allow reaction with water and filtering. A simple method of generating HCl gas for crystallization is to add concentrated sulfuric acid dropwise to concentrated HCl and pass the resulting gas through drierite (or similar water absorbing substance) prior to bubbling it through the solution of the base in ether or other solvent.

Diazomethane Org. Synth. Coll. Vol IV,250(1963)

CAUTION: Diazomethane is explosive (sharp edges trigger it) and poisonous, so avoid ground glass joints, chipped glassware, etc.

Fit a 125 ml distilling flask with a condenser and a long stem dropping funnel. The condenser is connected via an adapter to a 250 ml Erlenmeyer flask. Through the second hole in the stopper of the Erlenmeyer is placed an outlet tube bent so as to pass into and nearly to the bottom of a second Erlenmeyer which is not stoppered. Cool both receivers in an ice-salt mixture and place 10 ml ether in the first flask and 35 ml ether in the second flask. The inlet tube is below the surface of the ether in the second flask. Place 6 g KOH dissolved in 10 ml water, 35 ml Carbitol (monoethylether of diethylene glycol), a magnetic stirring bar and 10 ml ether in the distilling flask. Adjust the dropping funnel so that the stem is just above the surface of the solution in the distilling flask. Place a solution of 21.5 g p-tolylsulfonylmethyl-nitrosamide (org. Synth. C.V. IV,943(1963)) in 125 ml ether in dropping funnel. Heat the distilling flask in a water bath at 70-75° with magnetic stirring and add the nitrosamide at a regular rate over 15-20 minutes. Then add more ether from the dropping funnel at the previous rate until the distillate is colorless (50-100 ml ether). The distillate contains about 2.8 g diazomethane. If desired, the diazomethane may also be prepared in ethanol.

For the in situ generation of diazomethane see TL 1397(1973).

Spot Test for Some Psychedelics

Reagent and color formed

Compound	p-DMAB-TS	Ethanolic p-DMAB
LSD	Blue	Purple
Lysergic acid	Blue	Purple
DMT	Yellow	Purple
DET	Yellow	Purple
Bufotenine	Green	Blue
Ibogaine	Yellow-green	Green-blue
Psilocin	Brown	Deep blue
Psilocybin	Yellow-green	Purple
Ergonovine	Blue	Purple
Indole	Yellow	Orange
Tryptamine	Yellow-green	Purple
STP	Light yellow	—
Blank	Colorless	Light yellow

Use powders or residue of an extraction.

Add 2 drops of color reagent. Compare color produced with reference compounds on spot plate or other white porcelain surface.

Color Reagents:

p-DMAB-TS: To a cool soln of 65 ml H_2SO_4 in 35 ml H_2O, add 125 mg para-dimethylaminobenzaldehyde, dissolve, add 1-2 drops of $FeCl_3$-T.S. (9 g $FeCl_3$ in 100 ml H_2O).

Ethanolic p-DMAB: dissolve 2 g of para-dimethylaminobenzaldehyde in 50 ml ethanol, qs to 100 ml with HCL.

MISCELLANY
ON UNDERGROUND LABORATORIES

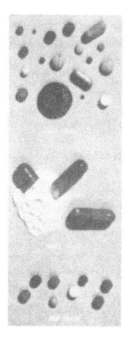

LSD. In 1971, about 274,000 LSD dosage units were seized; in 1972, 340,000; in 1973, 490,000. In the first four months of this year, already 364,000 units had been seized. It is estimated that only a fraction of the amount of LSD available is seized. Sixty-four sets of LSD punches are known to have produced LSD for the illicit market during 1973, by contrast with the previously recorded high of 27 sets of punches known to have been operative in 1971.

Cone-shaped LSD tablets were seized by BNDD agents in Detroit in June. These were the first LSD tablets encountered in this shape. The seizure consisted of 10,000 brick-red tablets. Analysis revealed approximately 40 micrograms of LSD per tablet. The method of packaging was also a first. The tablets were in plastic bags (1,000 per bag), folded in such a way as to form a cone. Each cone-shaped tablet measured six millimeters from base to apex, with the base slightly larger than the cone, giving a collared effect. The collar was 6.8 millimeters in diameter and 1.8 millimeters thick. On the collar face opposite the cone was a truncated cone approximately the same thickness as the collar. *(Nov. 1970)*

Late 60's European LSD on sugar cubes and candy (left) and mescaline (right).

Coca leaves and cocaine hydrochloride. *Above,* an unusual crystalline form; *below* cocaine as it usually appears on the illicit market.

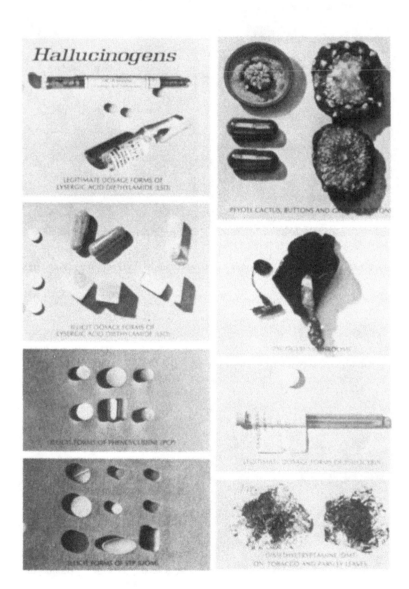

Hallucinogens

LEGITIMATE DOSAGE FORMS OF
LYSERGIC ACID DIETHYLAMIDE (LSD)

PEYOTE CACTUS, BUTTONS AND GROUND BUTTON

ILLICIT DOSAGE FORMS OF
LYSERGIC ACID DIETHYLAMIDE (LSD)

PSILOCYBE MUSHROOM

ILLICIT FORMS OF PHENCYCLIDINE (PCP)

LEGITIMATE DOSAGE FORMS OF PSILOCYBIN

ILLICIT FORMS OF STP (DOM)

DIMETHYLTRYPTAMINE (DMT)
ON TOBACCO AND PARSLEY LEAVES

The Perils of PCP

On January 16, 1974, a postal deliveryman on his rounds in Rockville, Maryland, found the owner of what purported to be an electronic circuit manufacturing plant lying unconscious in his laboratory.

The deliveryman called the local fire and rescue squad, who found on the premises quantities of ether, alcohol, and other inflammable substances. The rescue squad called the fire marshal. Upon examination of the laboratory, the marshal called DEA.

The presence of an illicit laboratory in the suburbs of the nation's capital (pictured on this page) had not gone entirely undetected. Federal authorities were first alerted to the possibility when a Wisconsin chemical firm informed DEA's Milwaukee office of orders received from the Rockville plant for precursors used to produce phencyclidine, a potent hallucinogenic drug, more commonly known as PCP.

A check with the Montgomery County police soon revealed that they, too, had the plant under investigation through a lead supplied to them by a laboratory apparatus distributor.

DEA and the county police were in the process of subpoenaing records from the two companies that had provided the leads when the call came from the fire marshal that brought the case to a sudden head.

The laboratory operator, when he woke up, was arrested, and is now awaiting trial in Montgomery County charged with the illegal manufacture of PCP. (As it turned out, he was found to be suffering from an overdose of a drug prescribed by his doctor.)

Seized on the premises were more than 90 pounds of PCP with a street value of $25 million—half of it in liquid form, the other half in powder form.

PCP in liquid form is sprayed on marijuana, parsley, oregano, or other plant leaves, and sold as "angel dust." In powder form it is sold under the deceptive name of "peace pills."

Developed as an animal tranquilizer by Parke-Davis & Co., a pharmaceutical firm in Detroit, it is now manufactured legally only by Bio-Ceutics, a firm in St. Joseph, Missouri. The drug is not prescribed for human beings because the range between effective and toxic doses is narrow.

Its unobtrusive but wide illicit diffusion is attributable in part to a reputation for being similar to—but safer than—LSD. Physicians, however often find PCP "trips" difficult to treat due to ignorance of the drug's identity; it is often sold to the unwary as mescaline, LSD, THC, or mixed with other drugs for illicit sale.

On March 1, six weeks after the Rockville seizure, a second major PCP ring closed down when DEA agents, after obtaining a search warrant, raided the attic of a small private home on the outskirts of Detroit, Michigan.

In the attic was an elaborate laboratory stocked with half a million dollars worth of drugs and equipped with an impressive array of test tubes, chemical filters, scales, and chemistry textbooks. Chemicals from the laboratory were pressed into pills at several other locations.

The entire drug ring, it was estimated, produced about a million PCP pills per week.

How many PCP users there are in the country is difficult to estimate. Its effects include feelings of weightlessness, diminishing body size, and catatonic rigidity, accompanied by a sense of dying or already being dead.

A physician at Detroit's Lafayette Clinic noted that the drug was so powerful his hospital was forced to discontinue tests of it when it could not find enough subjects who were willing to try it twice.

MDA LABORATORY FOUND IN DENVER

On May 24, 1974, a senior at the University of Colorado, was sentenced in the state's U.S. District Court to four years in prison and three years parole for manufacturing an hallucinogenic drug and possession for distribution.

His longtime partner, who was found guilty on the same two counts, received an indeterminate sentence under the Federal Youth Corrections Act not to exceed four years.

These were the largest producers on record of a drug called MDA—methylenedioxyamphetamine. Chemically related to both mescaline and amphetamine, MDA is a potent stimulant with a history of reported fatalities since it was first synthesized in the 1930's. Its hallucinogenic potency is reported to be three times that of mescaline.

The pair, who had been the subjects of recurrent investigations by federal enforcement officials, were spotted one morning in a passing car by a DEA agent working out of the sheriff's office in Boulder, Colorado.

The agent had, as it happened, been investigating their activities only last summer at an abandoned missile site 50 miles north of Cheyenne, Wyoming.

Previously, in 1971, they were believed to be operating a clandestine MDA laboratory in Eugene, Oregon, under the name of Northwestern Orangefield Oil and Aromatics Company.

DEA records also indicate that he had earlier been implicated in operations of clandestine laboratories in Los Angeles in 1966 and Boulder, Colorado, in 1967.

"They've popped up here and there for a few months at a time," said the undercover agent who recognized them. "But we never got enough evidence on them for a warrant."

He stopped them, he said, "to get ID's and to see what was going on."

In the possession of one of the suspects, he found a small caliber pistol as well as a large number of suspected amphetamines.

The two suspects were then taken to DEA's regional office in Denver. Both were later booked into Denver City Jail.

Meanwhile, the agents had found in the suspects' car records of a company called Solar Metallurgicals with a Denver address. The company, located in a large one-story cinderblock building, was unoccupied when they got there. Walking around the building, they noted a pervasive aroma of ether, although the windows were lined with tape.

They notified the Denver police, whose narcotics officers were, in fact, already in the process of investigating this company.

Late that afternoon a search warrant was executed by federal and local narcotics officers, resulting in the discovery of the largest illicit laboratory of its kind ever encountered by drug enforcement authorities.

They seized about 80 pounds of MDA in various stages of production. It is estimated that one ounce of MDA can be made into 14,000 dosage units, which sell on the illicit retail market for $3 each.

Washington, DC—DEA agents and Arlington County Police cooperated on a buy that netted a kilogram of cocaine and the arrest of four South American smugglers. The action began with an informant's phone call advising police that he had possible access to a large amount of cocaine. The drug, smuggled into the United States taped to the bodies of women couriers, was offered by a Peruvian identified as "Coco." The police requested DEA's help and together they set up two surveillance teams. After the informant had actually seen the cocaine in a suburban Washington motel room, he returned with a DEA undercover agent. The surveillance teams moved in once the substance was positively identified as cocaine. The kilogram cache was retrieved from its hiding place under a bed, and four Peruvian nationals were arrested. (1974)

Coos County, OR—A routine traffic check of a vehicle with Texas license plates revealed that it was registered in the name of a fugitive wanted by DEA. Surveillance of the vehicle resulted in issuance of two Oregon state search and seizure warrants. As a result of these warrants, nine defendants—four men and five women—were apprehended, including two DEA fugitives. A third fugitive, escaped during the search and is now being sought. Seized on the premises of the two searched residences were two well-equipped and operational methamphetamine laboratories and a total

ILLICIT LAB Seeing how dry the grass was, the landlord stopped to tell his tenant that he must water the lawn. When he knocked at the door he got no answer but smelled a pungent odor which he thought to be a dead body. Upon investigation, the Denver Police discovered the house contained an illicit STP and LSD laboratory. The entire water supply had been diverted to the laboratory.

of forty firearms, including two machine guns, a revolver with a silencer, a sawed-off shotgun, three pipe bombs, and an electronically detonated bomb. The presence of extensive "Weatherman" literature, in addition to the arsenal of weapons, has raised suspicions that the defendants may have been involved in a militant movement — a possibility now under investigation by the FBI. (1974)

Troy, Michigan — Special agents arrested seven suspects and seized 20 ounces of Phencyclidine (PCP), 85,000 LSD tablets, 75 pounds of marijuana and drug equipment. Simultaneously, agents ar-

rested the principal suspect, a Canadian, at Clinton Township, Michigan. The agents seized 2 ounces of PCP, 8000 baribiturate capsules and marijuana....**Washington, DC** — Agents arrested two men and seized about 8 ounces of cocaine.... **Yuma, Arizona** — Agents arrested a man and his wife when they delivered a pound of heroin to undercover agents. The suspects are Mexican nationals. A car was also seized.

INTERNATIONAL

JULY..**Luang Prabang, Laos**—GSI officers arrested a Lao national in possession of 11 pounds of opium...**Arica, Chile**—A joint investigation with Chilean Customs resulted in the arrest of a subject in possession of 4.4 pounds of cocaine and cocaine laboratory equipment...**Vancouver**—A joint DEA and Royal Canadian Mounted Police investigation culminated in the arrest of 18 suspects after the seizure of 800 pounds of hashish and 200 pounds of marijuana in a converted mine sweeper off the coast...**Columbia**—Colombian authorities arrested two subjects in possession of 20 pounds of coca paste...A cocaine laboratory near **La Vega** was raided and three defendants were arrested...Acting on information from DEA, Colombian agents seized a kilogram of cocaine and arrested a major violator...

JUNE...**Oquendo, Peru**—Agents from the Lima District Office assisted Peruvian authorities in an investigation which resulted in the seizure of a cocaine crystal laboratory. Four persons were arrested and small amounts of cocaine, coca paste and laboratory equipment were seized...**Nuevo, Laredo**—Special agents from Region 11 assisted Mexican Federal Judicial Police in the arrest of three Mexicans in possession of 24 pounds of cocaine and 2 pounds of brown heroin...**Santiago**—Chilean Police arrested three Chilean nationals and seized 2.2 pounds of cocaine...**Tocumen International Airport, Panama**—Panamanian Customs officials seized 4.5 pounds of cocaine and arrested three Mexican nationals.. ...**Okinawa**—In cooperation with the U.S. Air Force OSI and the U.S. Marine CID, special agents arrested two American civilians and one U.S. military man. An investigation led to the seizure of 4.4 pounds of heroin, a half ounce of liquid LSD and a quantity of "window pane" LSD...**Beirut**—Special agents assisted the Lebanese Judiciary Police in arresting two Syrian nationals and in seizing 27.5 pounds of opium...**Panama**—Two individuals were arrested by local authorities when they delivered 5.5 pounds of cocaine to undercover agents...**Ankara**—One man was arrested for selling 21 pounds of morphine base to a TNP undercover officer. He subsequently surrendered an additional 50 pounds of morphine base...**Milan**—The Italian National Police arrested a Chilean citizen at the Milan Airport in possession of 4.4 pounds of cocaine...

MAY...**Mexicali, Mexico**—Special agents assisted Mexican Federal Judicial Police in the arrest of one man in possession of 4 ounces of brown heroin...**Tijuana**—As three Mexican nationals delivered 20 ounces of brown heroin to an undercover agent, they were arrested by Mexican Federal Judicial Police...**Bogota**—Colombian authorities rounded up six Colombians and seized 20 ounces of cocaine and undetermined amounts coca leaves, chemicals and laboratory equipment...**Tijuana**—With agents acting undercover, the Mexican Federal Judicial Police seized 1,104 pounds of marijuana concealed in a load of watermelons on a truck from

Sinaloa...**Bombay**—Indian CBI Officers arrested three defendants and seized approximately 200 pounds of processed opium as it was delivered to an undercover special agent...**Santa Ana**—Special agents of the Nogales and Tucson District Offices assisted Mexican Federal Judicial Police in arresting one man in possession of approximately 3 pounds of brown heroin...**Tecate, Mexico**—Working in cooperation with agents from the San Diego District Office, Mexican Police arrested three defendants as they delivered 23 ounces of heroin to an undercover agent...**Popayan, Colombia**—Colombian authorities arrested one man and seized 9 pounds of cocaine...

APRIL...**Karachi, Pakistan** — Assisted by U.S. agents, Pakistani authorities completed an investigation that resulted in the arrest of five defendants and the confiscation of approximately 300 pounds of opium..**Santo Tomas, Mexico**—After placing a package containing 6 pounds of cocaine at a point on a clandestine airstrip, a man was arrested by Mexican Federal Judicial Police. As two men who were to take possession of the cocaine package arrived they also were arrested. Seized along with the cocaine were the Beechcraft airplane, which had been utilized to make the drop, and a van... **Vancouver, BC**—A four-month international investigation trailing from Canada to Tennessee to Switzerland climaxed in the arrest of two defendants, one a U.S. citizen. At the time of arrest the subjects were in the process of operating a clandestine MDA laboratory which was disguised as a perfume factory. During the search of the laboratory, 15 pounds of MDA powder, 300 pounds of MDA liquid, and enough chemicals and precursors to produce an estimated 2,000 pounds of MDA were seized..**Sonora, Mexico**—Special agents the Hermosillo District Office assisted the Mexican Federal Judicial Police in the search of passengers on a train originating in Sinaloa. A woman and her mother were arrested in possession of 4.5 pounds of brown heroin..**Laza, Bolivia**—Bolivian authorities assisted by U.S. agents seized a cocaine laboratory near Laza and two defendants were arrested..**Beirut**—Assisted by U.S. agents, the Lebanese Judiciary Police arrested one man as he delivered 5 pounds of tan heroin. (Fall 1973)

ROLE OF THE *DEA* LABORATORIES IN DRUG BUSTS

Evidence in drug investigations may exist in any form -- tablets, capsules, powders, liquids -- and can be present as large samples consisting of hundreds of containers, or as a minute residue.

The accurate analysis of these substances is the responsibility of the Drug Enforcement Administration's forensic laboratory system. This system consists of six regional laboratories -- in New York, Chicago, San Francisco, Dallas, and Washington, D.C. -- together with a special testing and research center in McLean, Virginia.

In the last fiscal year the DEA laboratories, under the direction of 120 professional chemists, analyzed 45,000 separate exhibits of drug evidence. In addition, they stand ready to offer assistance to any duly authorized law enforcement agency, at home or abroad, in laboratory planning, evidence handling, drug analysis, or court testimony at no cost to requesting agency.

Before going on to describe the specific services they perform in support of drug enforcement, I want to emphasize that they serve to supplement -- they do not supplant -- the services of state, county, and municipal laboratories.

The Forensic Chemist

The most important person in the DEA laboratory system is the forensic chemist. DEA requires the trainee chemist to undergo a special six-month training program. He is taught from the outset that each piece of evidence is unique and no step-by-step procedure can be written to cover all cases.

This training program is designed to acquaint him with new analytical procedures and to encourage independent thought, sound judgment, and ingenuity in the analysis of drug evidence. The new chemist is made aware of the importance of the chain of evidence, and the need for security in handling evidence, as well as his role as an expert witness.

Source Identification

A major concern of DEA special agents is the identification of the source of drug evidence. One of the methods they use to meet this requirement is through "ballistics examinations" of tablets. A ballistics examination, in the sense used here, is a combination of in-depth chemical analysis and tool-mark examination of tablets and capsules for the purpose of identifying the manufacturer of the drug. The technique consists of accurate measurements of size scoring, imprints, or bevels; the microscopic examination of the imperfections of the surface; the identification of all the components of the exhibit, primarily through microchemical tests; and comparison with exhibits contained in the Administration's National Authentic Drug Reference Library. These are exhibits obtained from drug firms and an accumulation of exhibits from clandestine manufacture.

Using this type of analysis, we are able to identify the manufacturing source of commercial dosage forms and to relate exhibits originating from common clandestine sources. In this way we are able to show diversion of legitimate products into illicit channels and to identify illegal distribution systems and marketing levels.

In the past year—to give you an example—the US has been flooded with small amphetamine tablets known as "minibennies." Ballistics examination indicates that most of these tablets originate from three sets of punches and dies used on multiple punch tableting machines. To date, over 30 million "minibennies" have been purchased or seized.

As with source identification of tablets and capsules through ballistics examination, we are also attempting to determine common sources of powers. We have developed methodology which will

provide intelligence on heroin evidence. This is not done routinely, but only on specific request. The technique has been used to trace distribution routes and to assist in the development of conspiracy cases. The technique consists of microscopic examination, the determination of elemental composition, and the determination of the ratios of heroin hydrochloride, morphine, 0^3 and 0^6 monoacetyl morphine, codeine, and acetyl codeine.

Special Analysis

Quantitative analysis of drug evidence is standard procedure in our laboratories. The nature of the charge and the sentencing of the defendant is, in many instances, dependent on the amount and purity of the drug. Large quantities and high purity usually indicate the higher echelon drug trafficker. Rapid, accurate, quantitative laboratory analysis can therefore be an aid to enforcement strategy.

In one conspiracy case, information was obtained that indicated a transfer of heroin had taken place in a garage. During the transfer one of the bags of

heroin had allegedly fallen and broken. Six months after the transaction had taken place the laboratory was asked to verify this information, even though the heroin had reportedly been swept up and there was no visible trace of it.

The laboratory devised a special vacuum pump and filters to take samples from various sections of the garage. The laboratory was able to confirm the presence of heroin, using rigorous isolation techniques combined with fluorimetry, thin-layer chromatography, and mass spectrometry. The laboratory's work played a major role in the successful prosecution of this case.

Recently, the compound 1, 3-dephenyl-2-methylaminopropane hydrochloride was identified in two different samples of methamphetamine powder. The identification was based on a combination of UV, IR, NMR and mass spectrometry. The presence of this compound in methamphetamine indicates that the clandestine operators were syn-

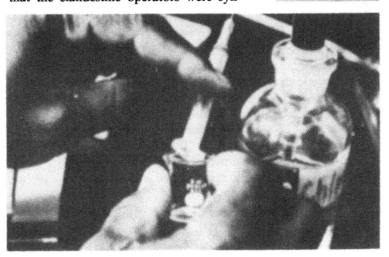

thesizing phenyl-2-propanone (P2P) from phenylacetic acid rather than purchasing this essential raw material. A byproduct of this synthesis is dibenzylketone, which, if not removed, would react when methamphetamine is made and produce 1, 3-diphenyl-2-methyl-aminopropane. This identification is very valuable intelligence for our agents who are looking for clandestine drug operators.

Methaqualone is being widely abused by youths in this country. Recently, in the northeast, some abusers were stricken with severe side effects, including bloody urine and gastrointestinal cramps. The tablets were found to be of clandestine manufacture. They contained fairly large amounts of impurities from a poor synthesis. The impurities were o-toluidine, o-aminobenzoic acid, and o-methyl acetanilide—again proving the old proverb "Let the buyer beware."

Clandestine Laboratories

In the past year, our forensic chemists participated with special agents in the seizure of over 30 domestic clandestine laboratories which were producing LSD, phencyclidine (PCP), dimethyl trypta-mine (DMT), methamphetamine, and liquid hashish. DEA chemists have also examined illegal laboratories in Europe, South America, and the Far East. These laboratories were producing heroin or cocaine.

In examining a seized illegal laboratory, we feel that the following information is best acquired by a chemist: production capability; estimated length of time in operation; source of chemicals and equipment; methods of waste removal; efforts at concealment; method of manufacture. This and other information can be used to measure the impact of an immobilized laboratory and can provide possible means to assist in the detection of similar laboratories.

Our chemists also have had the opportunity to participate with special agents in debriefing informants regarding the manufacturing of clandestine drugs. There is much literature on the commercial manufacture of most drugs of abuse. Clandestine manufacturers, however, do not always follow these procedures or use the same equipment and techniques. It is therefore necessary for a law enforcement agency to determine exactly how the drugs are made. This information has many investigative and intelligence ramifications.

Information From Drug Evidence

The laboratory is the focal point for almost all information regarding a purchase or seizure of drug evidence. This information must be retrieved, evaluated, and reported to the proper parties in a timely manner. Information retrieval and reporting mechanisms are therefore an important and integral part of our laboratory system.

Our laboratory data processing system is known as STRIDE (System to Retrieve Information from Drug Evidence). It is now in its third year of operation. In 1974, we will be installing in each laboratory computer terminals connected to the main computer in Washington. Every piece of evidence analyzed is entered into the system. The type of information includes: the subject's name; where the purchase or seizure was made; the amount of money expended; the suspected drug; what the drug was found to be; the purity of the drug; and what excipients and adulterants were present.

The system is designed to serve several purposes. It provides information for decisions as to whether or not a drug should be controlled by showing how much and in what form it is appearing in the illicit market. The "Early Warning System"—designed to make DEA aware of new abusable drugs—relies on STRIDE as one of its input sources.

Our Office of Intelligence also uses this system. By comparing potencies and adulterants in various exhibits, distribution systems can be identified, individuals can be tied together, and trends can be shown. Moreover, the use of a unique cutting agent for heroin such as methapyrilene, appearing in one section of the country, can provide investigative leads to our special agents.

The computerized laboratory analytical data of heroin exhibits is used to calculate the price per milligram of pure heroin. The price-purity data can be used by statisticians as a market index—like Dow Jones—to determine trends and changes in the domestic heroin market.

International Liaison

Drug abuse is an international problem. We exchange scientific information with police laboratories in over 60 countries, as well as the U.N. Laboratory and Interpol. Recently, the importance of our foreign liaison paid off when Canadian authorities brought to our attention 4-methoxyamphetamine, a new hallucinogen that was causing deaths. Shortly thereafter this drug made its appearance in the United States. The information supplied by Canadian authorities enabled DEA to bring this drug promptly under control.

Conclusion

The role of the forensic drug laboratory has changed over the last few years and is continuing to change. We must keep pace with these changes by using the most modern instruments and techniques. Demands are increasing from drug law enforcement officers for more information on exhibits for intelligence purposes. We must also be ready to respond to increasingly sharp questions from attorneys, who are today better prepared and more knowledgeable about drugs than ever before. The conclusions reached by the laboratory must leave no room for doubt since they are used within the criminal justice system. The forensic laboratory, in the final analysis, is an instrument of justice.

DEA WATCHED LIST OF CHEMICALS

(Compiled by Richard C. Hall III)

Many of these chemicals are suspicious only if ordered by individuals, in large or in repeated orders. Others are suspect under any conditions.

Acetic Anhydride
Acetone Dicarboxylic Acid
N-Acetylanthranilic Acid
N-Acetylmescaline
Aluminum Foil (?)
Ammonium Acetate
Anthranilic Acid
Barbituric Acid
Benzaldehyde
Benzocaine
Benzyl Methyl Ketone
Beta Keto Glutaric Acid
mono-Bromobenzene
Butacaine
Chloroamphetamine
o-Chloroaniline
Chloromethanamphetamine
Citral
Cyclohexanone
Dextrose, anhydrous
Dextrose, monohydrate
5,5-Dibromobarbituric Acid
2,5-Diethoxytetrahydrofuran
Diethylamine
Diethylamine Malonate
1,3-Diethyl-2-Thiobarbituric Acid
Dihydroergotamine
Dimethylamine
Dimethyl Acetone

Dimethyl Beta Keto Glutaric Acid
Dimethyl Sulfoxide (DMSO)
Ephedrine
Ergocristine
Ergocryptine
Ergonovine
Ergotamine
Ergothioneine
Ethyl Ether
Formamide
Formic Acid
Glyoxylic Acid
Hexane
Hydrogen
Hydrogen Iodide
Hydroxylamine
Hydroxylamine Hydrochloride
5-Hydroxyuracil
Indole
Isobarbituric Acid
Lactose
Lidocaine
Lithium Aluminum Hydride
Magnesium Metal (esp. Turnings)
Malonic Acid
Methylamine
Methylamine Hydrochloride

2-Methyl Benzoxazone
3,4-Methylenedioxy
    compounds (any)
N-Methylmescaline
N-Methyl Formamide
Monoethanolamine
5-Nitrobarbituric Acid
Nitroethane
5-Nitrosothiobarbituric Acid
Olivetol
Oxalyl Chloride
Phenyl-2-Propanone
Phenylacetic Acid
Phenylacetone
Phenyl Magnesium Bromide
Phenylpropanolamine
Phosphorus Oxychloride
Piperidine
Piperonyl & related
    compounds
Potassium Cyanide
Procaine
Quinine
Sodium Acetate
Sodium Cyanide
Sodium Formate
Sulfanilamide
Thiourea
Toluene
p-Toluene Sulfonic Acid
o-Toluidine
Toluol
Trifluoracetic Acid
2,4,5-Trihydroxypyrimidine
3,4,5-Trimethoxyben-
    zaldehyde
Tryptophan
Tyrosine
Urea